Principles Of EKG Interpretation

by

FREDERICK J. CONDO

Bloomington, IN  Milton Keynes, UK

First published by AuthorHouse 3/16/07

ISBN: 978-1-4343-0502-2 (sc)

Printed in the United States of America
Bloomington, Indiana

This book is printed on acid-free paper.

PRINCIPLES OF EKG INTERPRETATION
CONTENTS

INTRODUCTION

TO PRINCIPLES OF EKG INTERPRETATION

The purpose of this manual is to present a method for acquiring skill in interpretation of the electrocardiogram. It should be used as a learning tool and not as a reference text. The manual is designed to carry the totally uninitiated student as well as the practicing physician to an understanding of spatial electrocardiography.

Among those who should be able to read EKGs include residents, interns, medical students, nurses in the emergency department and ICU/CCU, nurse practitioners, nurse anesthetists, physician assistants and personnel involved in pre-hospital care.

Spatial analysis of the electrocardiogram was developed by Robert P. Grant, M.D. who taught it between 1947 and 1950 at Emory University School of Medicine, Atlanta, Georgia.

A technique of spatial analysis is presented in a systematic fashion in Chapter 2. It was taken from *Spatial Analysis of the Electrocardiogram, Chapter 1*, by Irwin Hoffman, M.D. Dr. Hoffman was my mentor in electrocardiography from 1970 to 1974 during my residency in medicine and fellowship in cardiology at Cedars-Sinai Medical Center, Los Angeles, California.

In order to stimulate the imagination of the reader, pertinent illustrations and concepts from the following publications have been utilized in the manual.

1. *Clinical Electrocardiography*, The Spatial Vector Approach, Robert P. Grant, M.D., McGraw-Hill Book Company, 1957

2. *Introduction to Electrocardiography*, 2nd edition, J. Willis Hurst & Robert J. Myerburg, McGraw-Hill Book Company, 1973

3. *Diagnostic Electrocardiography & Vectorcardiography,* H. Harold Friedman, M.D., FACP, FACC, McGraw-Hill Book Company, 1971

4. *Clinical Vectorcardiography*, 2nd edition, Te-Chuan Chow, M.D., Robert A. Helm, M.D., & Samuel Kaplan, M.D., Grune & Stratton, 1974

5. *The Hemiblocks,* Mauricio B. Rosenbaum, M.D., Marcelo V. Elizari, M.D., Julio O. Lazzari, M.D., Tampa Tracings, 1970

6. *The Disorders of Cardiac Rhythm*, Leo Schamroth, Blackwell Scientific Publications, 1971

7. *The Cardiac Rhythms*, Raymond E. Phillips, M.D., F.A.C.P. & Mary K. Feeney, R.N., B.S.N., W.B. Saunders Company, 1973

8. *Electrocardiography*, Third Edition, E. Grey Dimond, M.D., Paul Schlesinger, M.D., Rafael L. Luna, M.D., The Corinth Press, 1961

9. *Structure And Function Of The Cardiovascular System*, Second Edition, Robert F. Rushmer, M.D., W.B. Saunders Company, 1976

10. *Practical Electrocardiography,* Fourth Edition, Henry J.L. Marriott, M.D., The Williams & Wilkins Co., 1968

Pattern interpretation of the electrocardiogram is the most commonly employed method in clinical practice. Most clinicians that use the pattern method get excellent results. However, there are those who feel that spatial analysis should be made available to anyone desiring to enhance his or her understanding of the differences in electrical activity between normal and diseased hearts.

Both pattern and spatial vector approaches have advantages and disadvantages associated with their use. Inherent with pattern recognition is the large number of configurations that must be committed to memory. Spatial analysis unfortunately often carries the stigma of "vector-phobia".

An article by J.W. Hurst, M.D. (co-author of *Introduction to Electrocardiography*, 2nd edition) entitled, "Methods Used to Interpret the 12-Lead Electrocardiogram: Pattern Memorization versus the Use of Vector Concepts", appeared in *Clinical Cardiology*, Vol. 23, No. 1: p. 4-13. In the article, Dr. Hurst indicates his support of spatial analysis of the electrocardiogram. The summary of the article is reprinted below.

"This article extols the value of using Grant's approach to the interpretation of electrocardiograms (ECGs). The essay includes a discussion on how people learn and emphasizes the difference in memorizing information, thinking, and learning. Simply stated, the brains of most people are not designed to memorize countless numbers of ECG patterns. Accordingly, the essay supports the view that a method of interpretation must be used, and the reader is encouraged to use basic principles of electrocardiography, including vector concepts, to interpret each ECG."

Spatial analysis virtually eliminates the need to memorize literally hundreds of patterns. A relatively small number of spatial vector profiles describe most of the electrocardiographic abnormalities commonly encountered in clinical practice.

Pattern interpretation may be likened to attempting to identify an individual from a verbal description. This is often very difficult. The spatial vector method, on the other hand, is similar to viewing a photograph or "mug shot" of a person often resulting in an almost immediate recognition. Word descriptions are analogous to electrocardiographic patterns, whereas, photographs are comparable to spatial "profiles".

The following is reprinted from the Preface of *Clinical Electrocardiography*, The Spatial Vector Approach. It conveys how Dr. Grant felt about patterns and vectors.

"It is not meant by this that vector methods should supplant the more familiar "pattern" methods of interpretation, but rather that they should supplement them. From the clinical point of view, when a tracing has the classic pattern of acute myocardial infarction, it is no more necessary to convert it into vectors than it is necessary to get an accurate measurement of body temperature when the patient has an obvious raging fever. However, when the tracing is perplexing or borderline, or when there is a slight difference in a follow-up tracing which is difficult to evaluate, then the vector method is the most accurate, objective, and rational method for interpretation that is so far available".

Willem Einthoven, M.D., a Dutch electrophysiologist, introduced in 1901 the standard bipolar limb leads I, II, and III. He determined the location of the positive and negative electrodes of each of these limb leads. In this manual, an electrocardiogram will be referred to as an "EKG" in Einthoven's honor. The three letters "EKG" are taken from the Dutch word "**E**lektro**K**ardio**G**ram".

Frank N. Wilson, M.D. introduced the unipolar limb leads VR, VL, and VF in 1932. The deflections produced by these leads were very small and difficult to read. Goldberger eliminated this problem by rearranging the electrical circuitry in a way that resulted in deflections one and one half times larger with little or no distortion. The letter "a", standing for "augmented", was added to VR, VL, and VF creating aVR, aVL and aVF.

Bayley constructed the frontal plane triaxial reference figure by moving the axes of the standard limb leads so that they each passed through a common central point the null or "e" point of the heart. By combining the triaxial system of Bayley with the unipolar limb lead axes, the hexaxial reference figure was produced. Please refer to the illustrations of the reference figures in the pages that follow.

After completion of this manual, the reader should be:

1) Knowledgeable of EKG lead systems and their relationship to left-right, superior-inferior, and anterior-posterior axes of the thorax.

2) Able to determine the *mean P wave, QRS, ST segment, and T wave vectors in the frontal and transverse planes.*

3) *Able to analyze spatially the electrocardiograms of children and adults.*

4) Able to use the "2-3-4 Rule" to differentiate between normal and pathologic initial QRS forces, i.e., the Q wave in lead I and lead aVF and the R wave in lead V_1 or V_2.

5) Able to identify a normal EKG, right ventricular hypertrophy, left ventricular hypertrophy, right bundle branch block, left bundle branch block, left anterior hemiblock, left posterior hemiblock, myocardial infarction, and ventricular pre-excitation.

CHAPTER ONE:
BASIC CONCEPTS

A. The Recorder

The modern electrocardiographic recorder (EKG machine) is designed to perform the following functions:

a. Measure electrical voltage (forces) generated by the heart

b. Measure the time it takes to depolarize and repolarize the heart

c. Determine in what direction forces generated by the heart are moving.

The EKG machine is, therefore, a voltmeter, timepiece, and direction finder

1. Voltmeter

A major component of the EKG machine is the galvanometer. This device, made up of an indicator and positive and negative electrodes, measures voltage. If the positive and negative electrodes of the galvanometer are attached to the skin at two different points, electrical energy generated by the heart will be detected by the meter. Since cardiac potentials recorded from the skin rarely exceed 2 millivolts (mV), amplification is necessary. A flow of current through the galvanometer causes its dial to move.

An illustration of a galvanometer at rest is shown in Figure 1. Its dial reads zero. If an electrical force of four mV is applied to the positive wire of the galvanometer, the dial will register +4 mV (Figure 2). On the other hand, if an electrical force of four mV is applied to the negative wire of the galvanometer, the dial will register –4 mV (Figure 3).

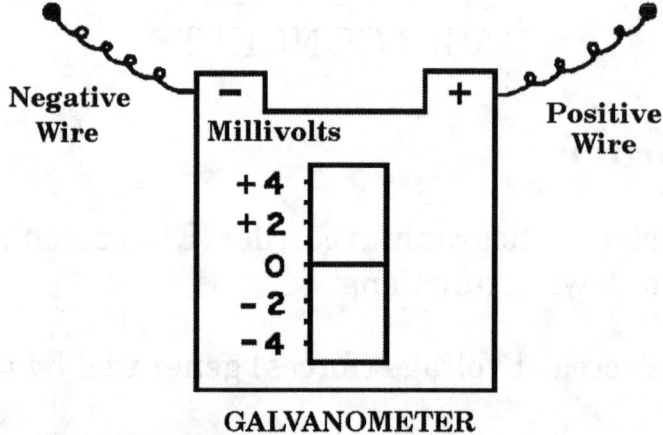

Negative Wire

Positive Wire

Figure 1: Zero voltage

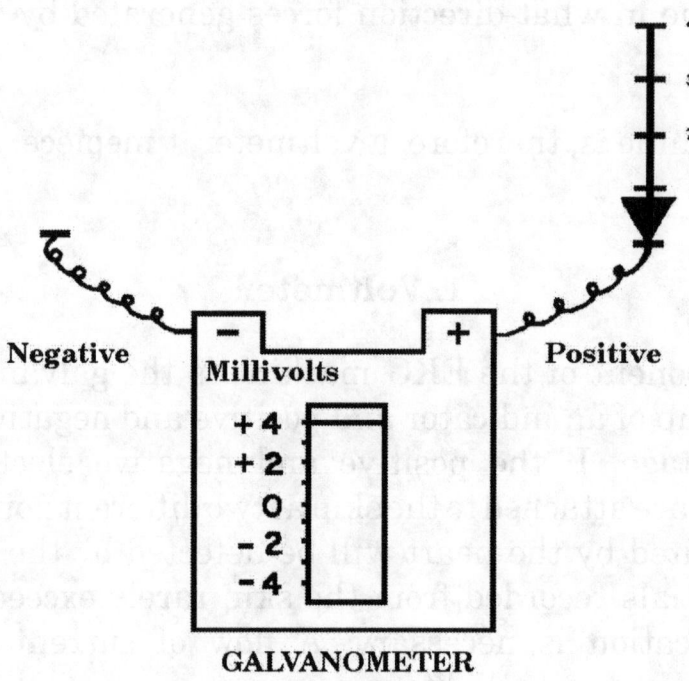

Negative

Positive

Figure 2: Positive 4 mV

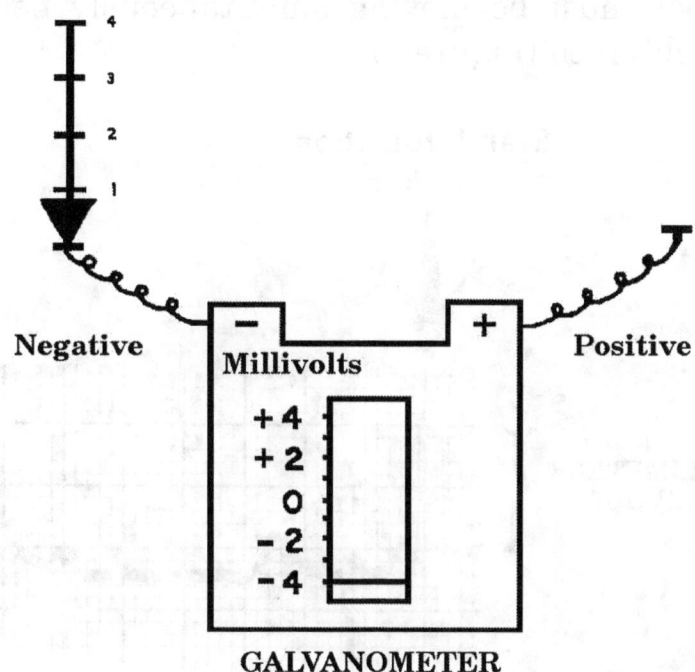

Figure 3: Negative 4 mV

The distance the dial moves is dependent on the amount of voltage generated by the heart muscle. The greater the voltage, the farther the dial will move.

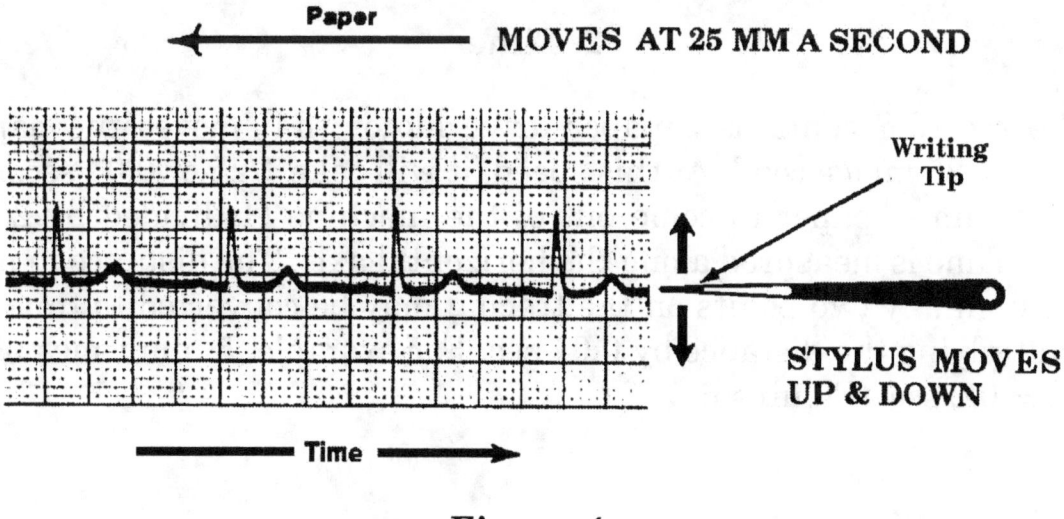

Figure 4

If a stylus is attached to the dial and brought in contact with *moving* EKG paper, a tracing will be inscribed. However, both stylus and

3

recording paper must be moving simultaneously before a legible record can be obtained (Figure 4).

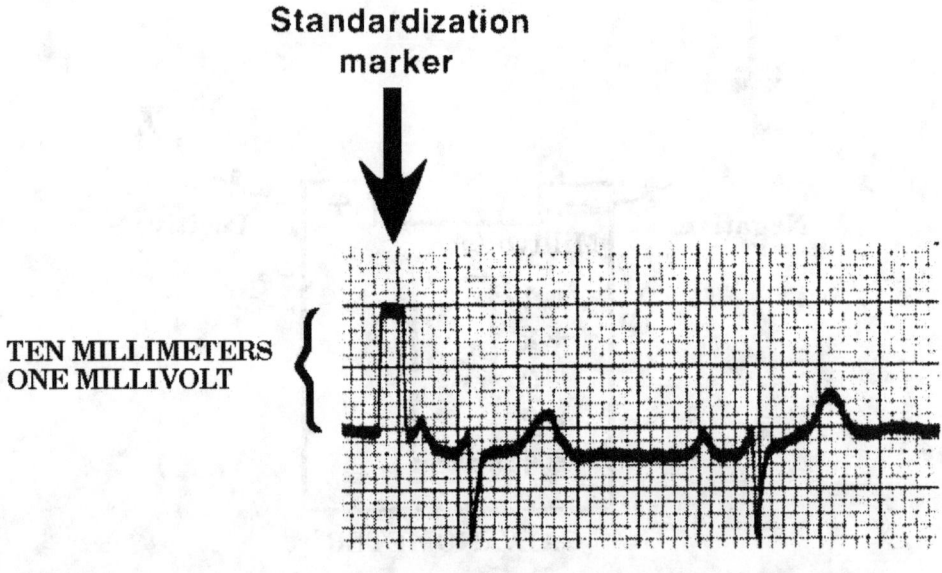

Standardization marker

TEN MILLIMETERS
ONE MILLIVOLT

Figure 5

A calibration device in the recorder (manual or automatic) is used to adjust the sensitivity of the galvanometer so that the introduction of a 1 mV signal will produce a deflection of 10 mm (Figure 5).

2. Timer

The recorder contains a paper drive motor that moves the paper exactly 25 mm/second. At this speed, it will take 1/25 or 0.04 second, for 1 mm of paper to come out of the machine (1 divided by 25 = .04). Time is measured along the horizontal axis. The distance in mm between any two points on the tracing may be converted to time by multiplying the distance by 0.04 sec. For example, 5 mm times 0.04 sec = 0.2 sec. (Figures 6 & 7)

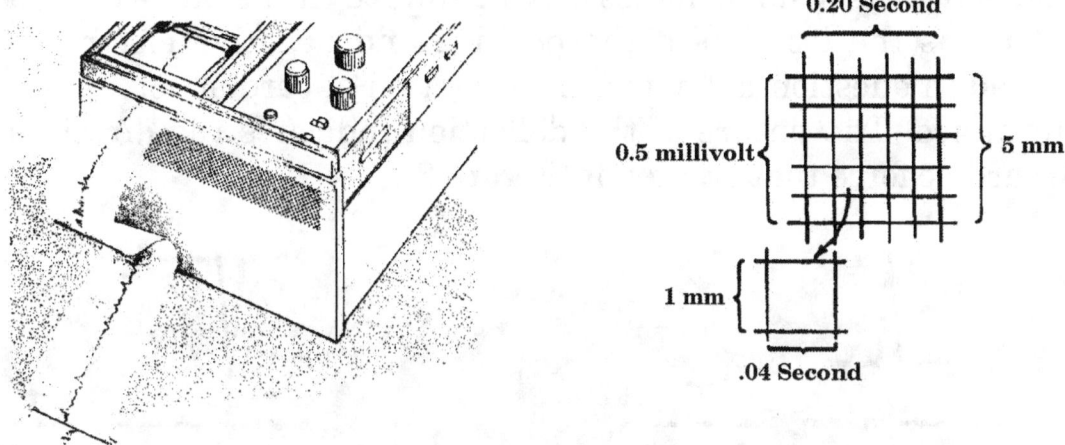

Figure 6

Conventional EKG paper is crosshatched by horizontal and vertical lines drawn 1 mm apart, forming a grid. Voltage is measured along the vertical axis (Figure 7).

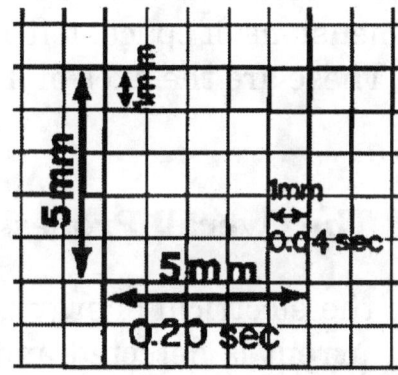

Figure 7: EKG paper enlarged

3. Direction Finder

The recorder is designed so that when forces generated by the heart are moving toward the positive electrode of a lead, a positive deflection will be recorded; when forces are moving toward the negative electrode of a lead, a negative deflection will be recorded.

Each lead system has a particular location for its positive and negative electrodes. A positive deflection in any lead is the result of forces moving toward the positive electrode of that lead. A negative deflection is the result of forces moving toward the negative electrode. As long as the location of the positive and negative electrodes of the lead in question is known, the direction of cardiac forces can be determined. The nature of the deflection indicates the direction of the cardiac force that created it (Figure 8).

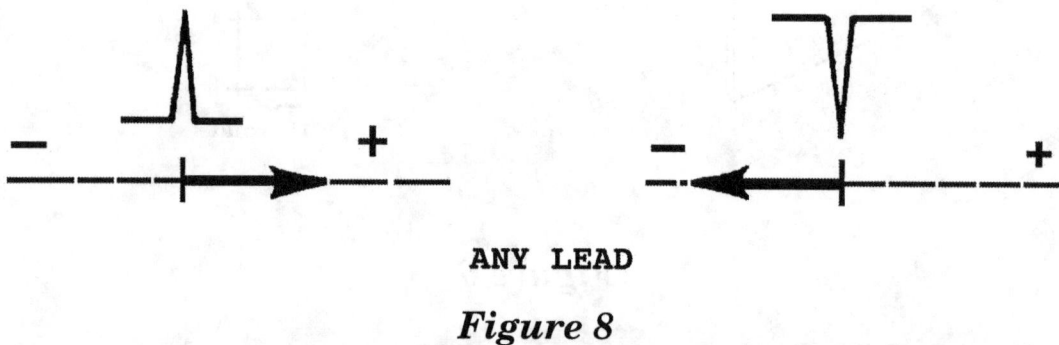

ANY LEAD

Figure 8

An EKG appears to be a two-dimensional graph. Actually, it contains three-dimensional information of the electrical activity of the heart. This is so because the EKG records from lead systems whose axes are parallel to the three dimensions of space: left-right, superior-inferior and anterior-posterior. These are the axes of I, aVF, and V_2, and will be discussed later.

4. The Overall Process

Only a small portion of the electrical energy generated by the heart reaches the skin. This current is detected and then made larger by an amplifier contained in the recorder. The tracing that is obtained is referred to as an electrocardiogram (Figures 9, 10).

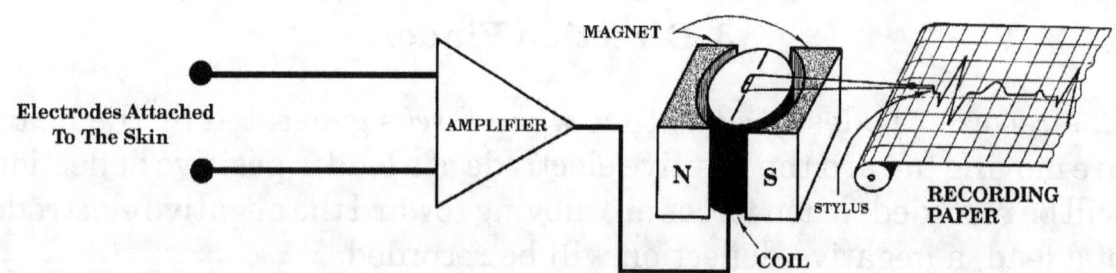

Figure 9: Recorder Schematic

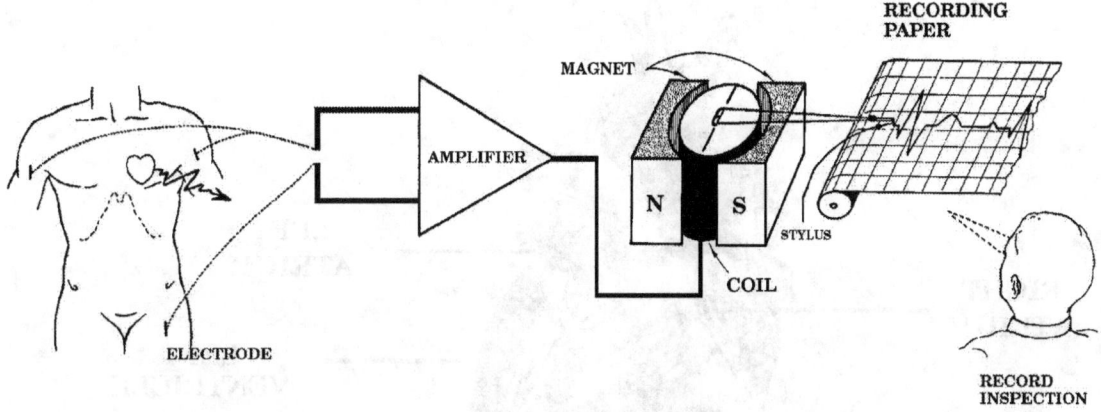

Figure 10: Overview of Process

B. Anatomy of the Heart

1. Chamber Location

The thorax will be treated as a cylinder in this manual. Within the thoracic cylinder, the right ventricle is located anterior and to the right, and the left ventricle posterior and to the left. The interventricular septum that separates the two ventricles is anatomically part of the left ventricle and lies almost parallel to the frontal plane. (Figures 11, 12, 13).

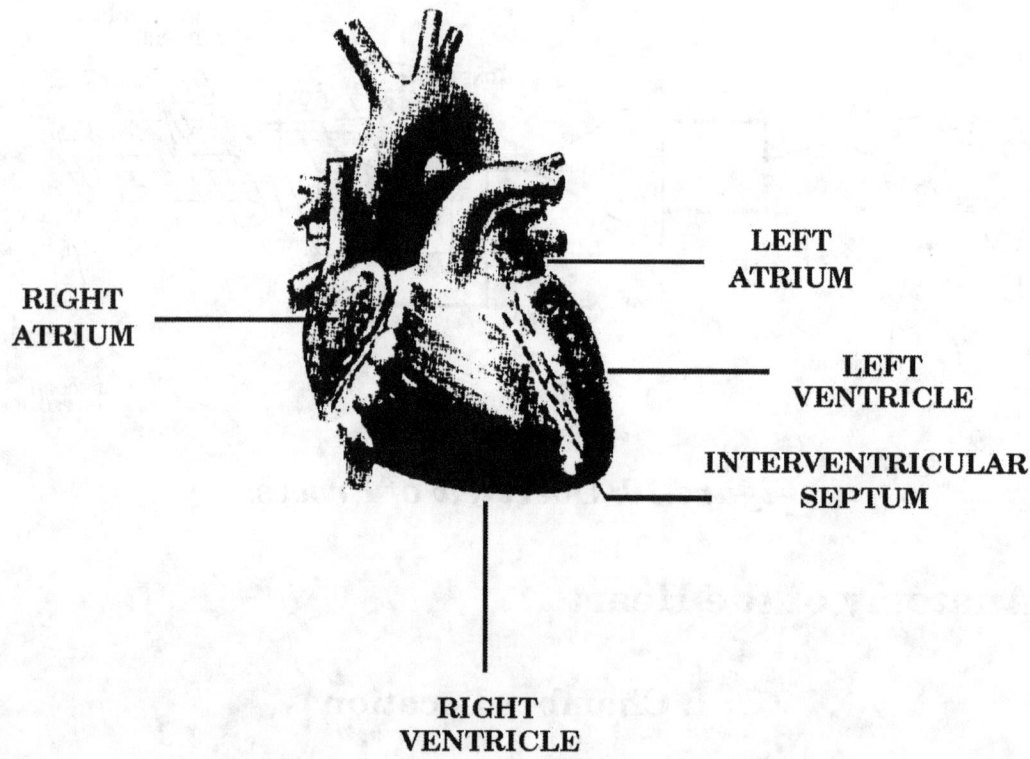

RIGHT
ATRIUM

LEFT
ATRIUM

LEFT
VENTRICLE

INTERVENTRICULAR
SEPTUM

RIGHT
VENTRICLE

Figure 11: Anterior View of the Heart

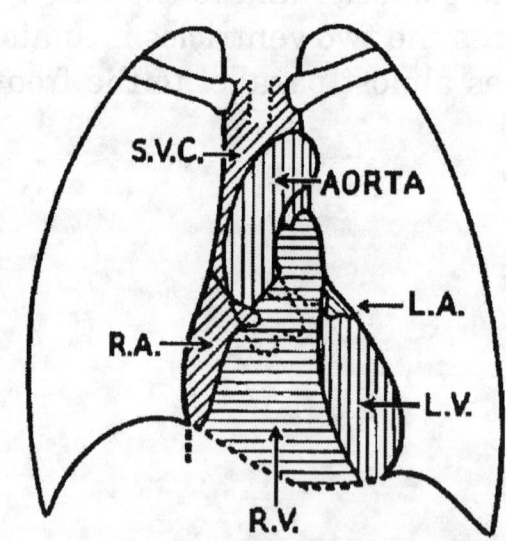

Figure 12: Anatomy as Seen on Chest X-ray

POSTERIOR

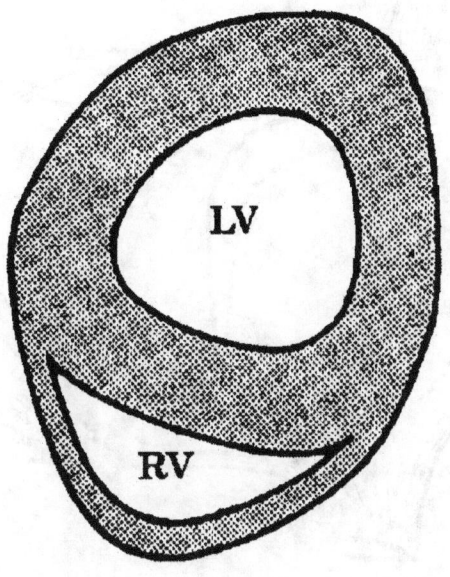

ANTERIOR

Figure 13: Cross Section

2. Conduction System

All myocardial cells have the property of conductivity, i.e., the ability to conduct the cardiac impulse at a given velocity to responsive adjacent tissue. In some cells, conductivity is more highly developed than in others. Specialized tissue capable of rapid transmission of impulses is known collectively as the "conducting system". It is made up of the S-A node, intra-atrial tracts, internodal tracts, A-V node, bundle of His, right bundle branch, and the left bundle brand with its anterior and posterior divisions.

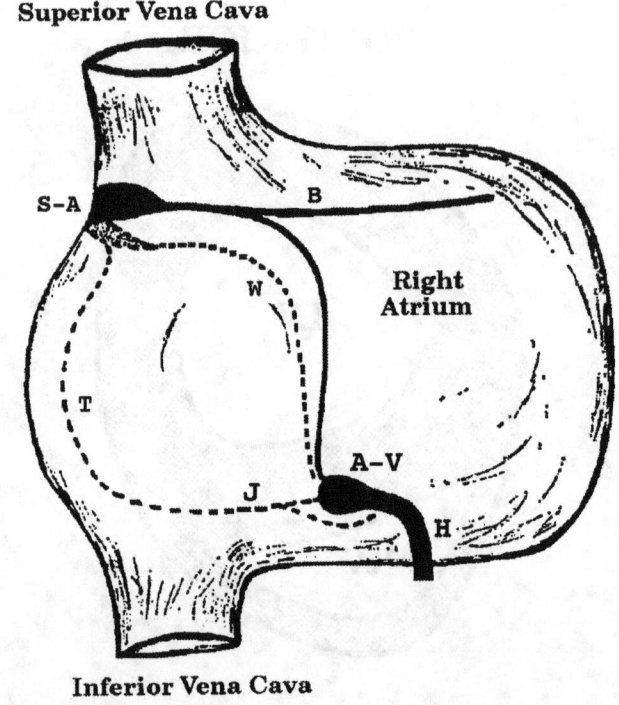

Superior Vena Cava

S-A

B

Right Atrium

W

T

A-V

J

H

Inferior Vena Cava

Figure 14: Side View of the Right Atrium

A diagram illustrating the intra-atrial and internodal pathways is shown in Figure 14:

   a. SA is the S-A node, the pacemaker of the heart

   b. B is Bachman's bundle which connects the right and left atria

   c. W is Wenckebach's bundle, an internodal tract

   d. T is Thorel's pathway, an internodal tract

   e. A-V is the A-V node

   f. J is James tract which bypasses the A-V node

   g. H is the bundle of His

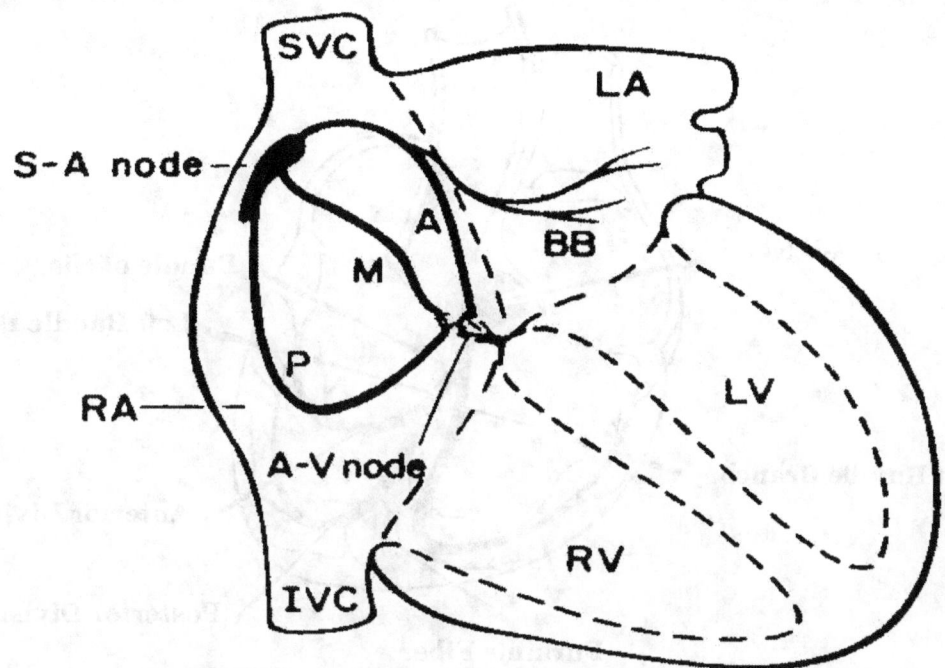

Figure 15: Atrial Conduction Pathways

Figure 15 is a diagram of the heart illustrating:

    a. A, the anterior internodal pathway

    b. M, the middle internodal pathway

    c. P, the posterior internodal pathway

The anterior, middle, and posterior internodal pathways connect the SA node with the AV node.

    d. BB, Bachmann's bundle connects the right and the left atria.

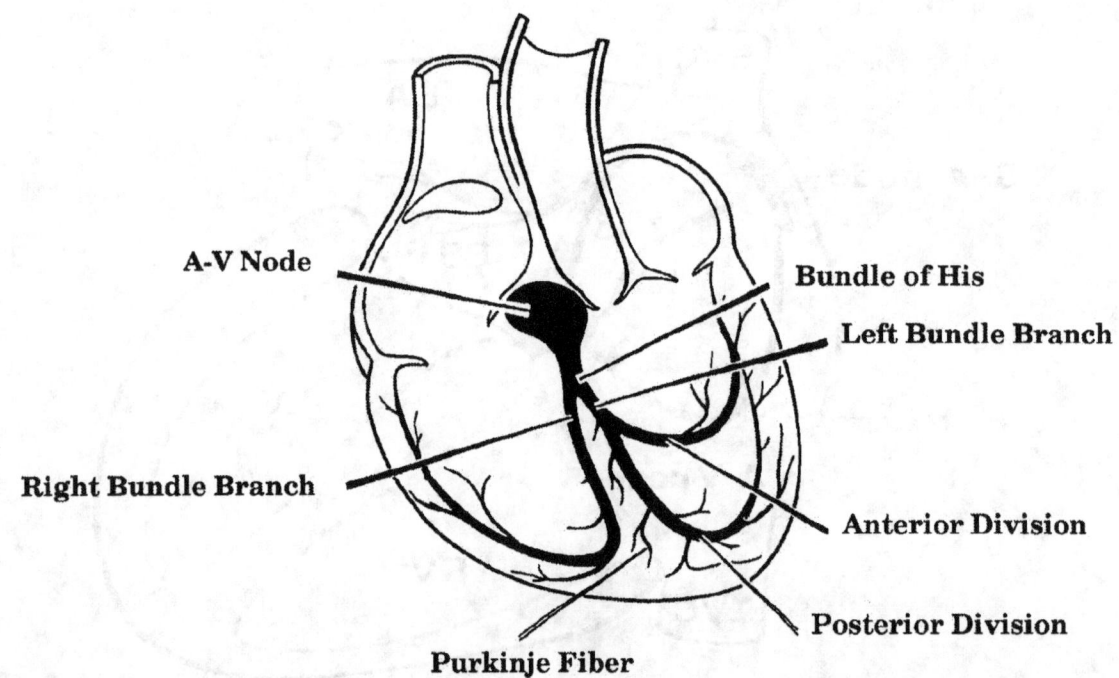

A-V Node

Bundle of His

Left Bundle Branch

Right Bundle Branch

Anterior Division

Posterior Division

Purkinje Fiber

Figure 16

The ventricular conduction system is made up of three anatomically distinct fascicles:

    a. The right bundle branch

    b. The anterior division of the left bundle branch

    c. The posterior division of the left bundle branch

Familiarity with the trifascicular configuration of the ventricular conduction system is essential. The conduction system will be discussed in Chapter 5.

C. Nomenclature And Intervals

The letters P, Q, R, S, T and U were chosen at random by Einthoven to identify the deflections produced by the electrical activity of the heart.

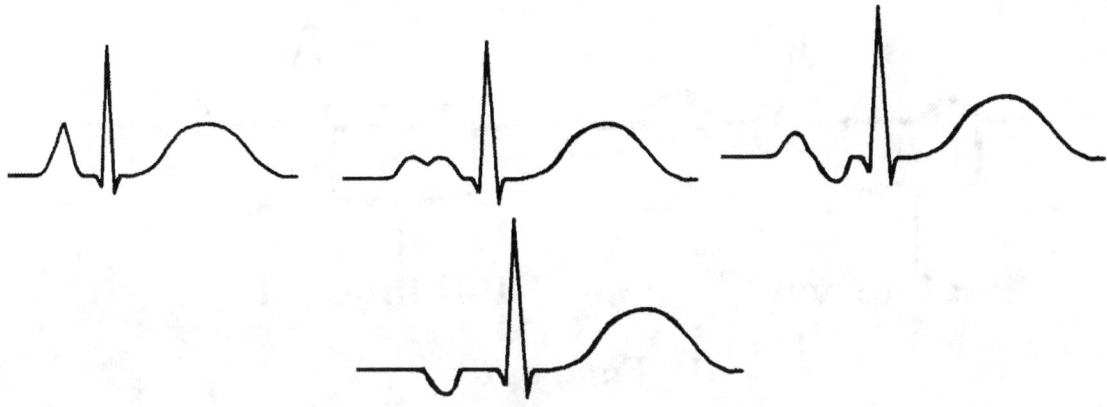

Figure 17: P Wave Shapes

1. P Wave

The P wave is the graphic representation of atrial depolarization. P waves are to the atria what QRS complexes are to the ventricles. P waves could be thought of as the atrial "QRS". Just like QRS complexes, P waves may be narrow, wide, notched, upright, biphasic or negative. The duration of the P wave is the time required for atrial depolarization (Figure 17).

2. PR Interval

The PR interval represents the time required for a stimulus to move from the SA node through the AV node to the Bundle of His. It is measured from the beginning of the P wave to the beginning of the QRS complex regardless of whether the initial QRS deflection is a Q wave or R wave. At a normal heart rate the PR interval is between 0.12 and 0.20 sec. Prolongation of the PR interval may be due to a wider than normal P wave (intra-atrial block), a delay at the AV node (AV block), or block in the Bundle of His. (Figure 18)

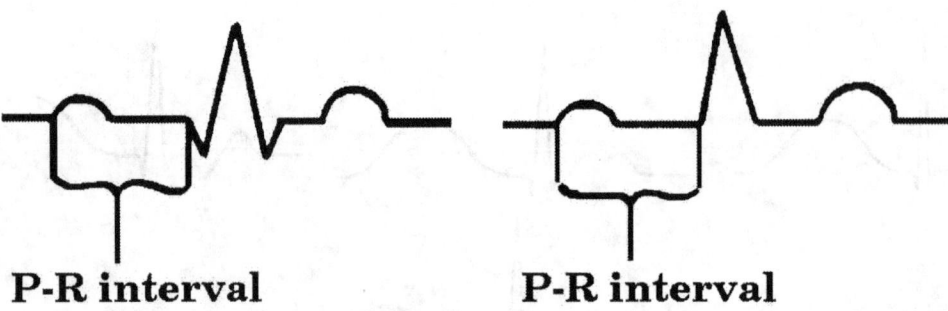

P-R interval   **P-R interval**

Figure 18

3. QRS Complex

The QRS complex is the graphic representation of ventricular depolarization. The illustration in Figure 19 is enlarged and demonstrates the classic QRS configuration. The maximum normal duration of the QRS interval is 0.08 sec. in children under 5 years of age, 0.09 sec. in children from 5 to 14 years of age, and 0.10 sec. in older children and adults.

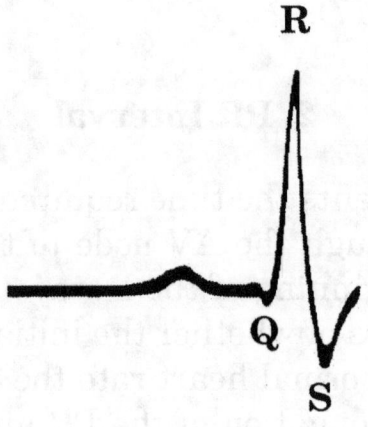

Figure 19

a. Q wave – *initial* downward or negative deflection followed by an R wave.

b. R wave – the first upward or positive deflection whether preceded by a Q wave or not.

c. S wave – downward or negative deflection following an R wave.

d. R' wave (R prime) – a second upward or positive deflection.

e. S' wave (S prime) – downward deflection following an R' wave.

f. QS complex – single downward deflection representing the entire QRS.

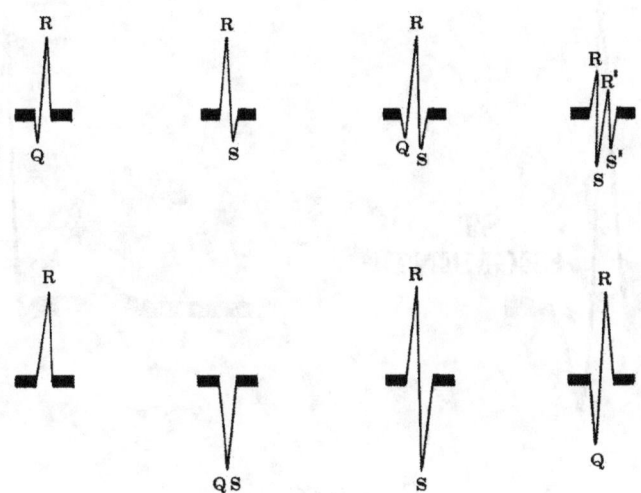

Figure 20: Common QRS Configurations

4. QT interval

The QT interval is measured from the beginning of the QRS complex to the end of the T wave. The normal range in adults is 0.30 to 0.45 sec. Digitalis, hypercalcemia, and increases in heart rate shorten it. Acute myocardial infarction, hypokalemia, and hypocalcemia prolong it (Figure 21).

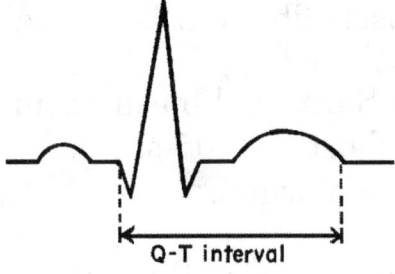

Figure 21

15

5. ST Segment and T Wave

The ST segment represents early ventricular repolarization and the T wave late ventricular repolarization. A complete cardiac cycle is shown below (Figure 22).

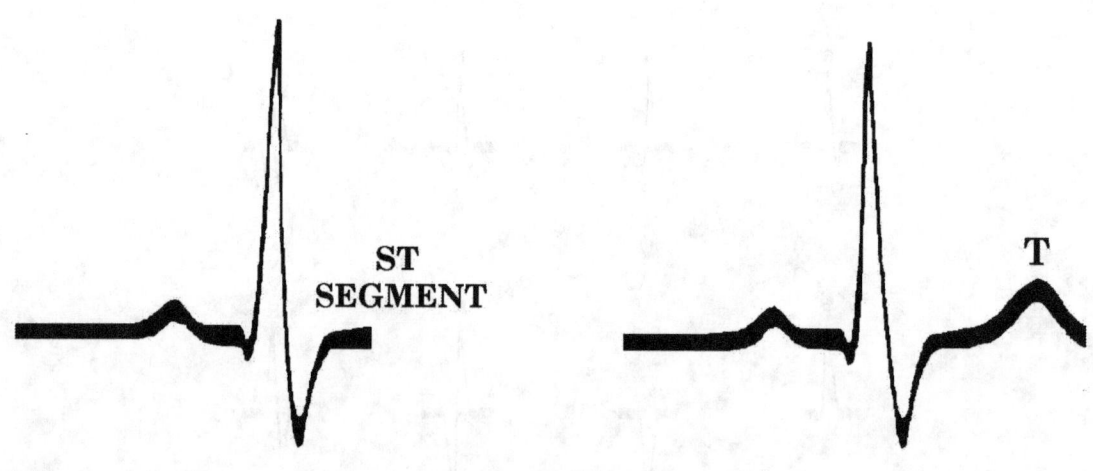

Figure 22

6. Interval Summary

Figure 23 illustrates the normal values for the important intervals used in electrocardiography. These are average measurements only.

The P wave, the first deflection of the electrocardiogram, represents the time for atrial activation. It should not exceed 0.11 sec. in duration.

The PR interval represents the time for an impulse to go from the SA node to ventricular muscle fibers. It is normally 0.12 to 0.20 sec.

The duration of the QRS interval has a normal range of 0.06 to 0.10 sec. QRS intervals as short as 0.05 sec. and as long as 0.11 sec are considered by many to be normal.

The QT interval is measured from the beginning of the QRS complex to the end of the T wave. It is lengthened in congestive heart failure, myocardial infarction, torsades de pointes, and hypocalcemia.

Quinidine and procaineamide also prolong the QT interval. It is shortened by digitalis, calcium excess and hyperkalemia.

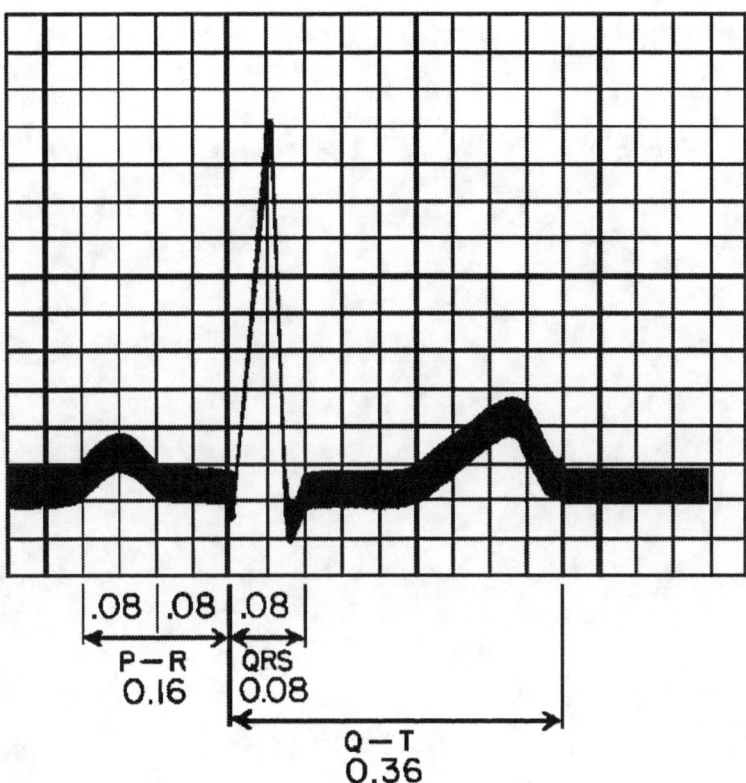

Figure 23

18

CHAPTER TWO:
VECTORS AND LEADS

A. General Considerations

Mechanical and electrical forces possess magnitude and direction. By definition they are vectors and may be represented by an arrow. The length of the arrow represents magnitude and its inclination indicates direction. A third attribute, the "sense" of the vector, is defined by the location of the arrowhead, which, by convention, is designated positive.

It is difficult to render the proper three-dimensional perspective of an arrow on a two-dimensional piece of paper. Spatial illusions may be obtained, however, using various shapes, shading and angles. The technique is illustrated in Figure 1.

AWAY FROM PATIENT   TO PATIENT'S LEFT   TO PATIENT'S LEFT
TOWARD READER   AWAY FROM READER   TOWARD READER

Figure 1

1. Mechanical Vectors

The illustration in Figure 2 shows a hand pushing against a wall. The force exerted against the wall may be expressed as an arrow, the symbol for a vector. The arrowhead indicates the direction of the force and the length of the arrow represents the magnitude of the force applied (Figures 3 & 4).

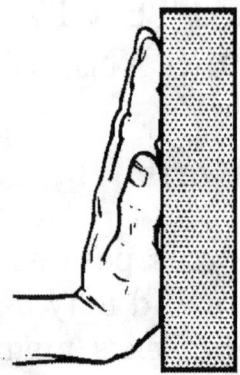

Figure 2

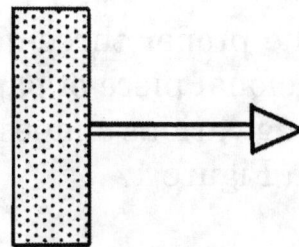

Figure 3

A pulling of the wall may be expressed by reversing the direction of the arrow (Figure 4).

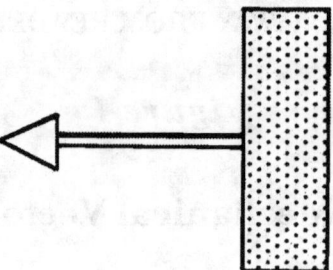

Figure 4

In Figure 5, a scale with weighing pans and dial is illustrated. A four-pound weight has been placed on the right weighing pan causing the dial to move to +4 pounds. This weight represents a mechanical force of four pounds pushing downward.

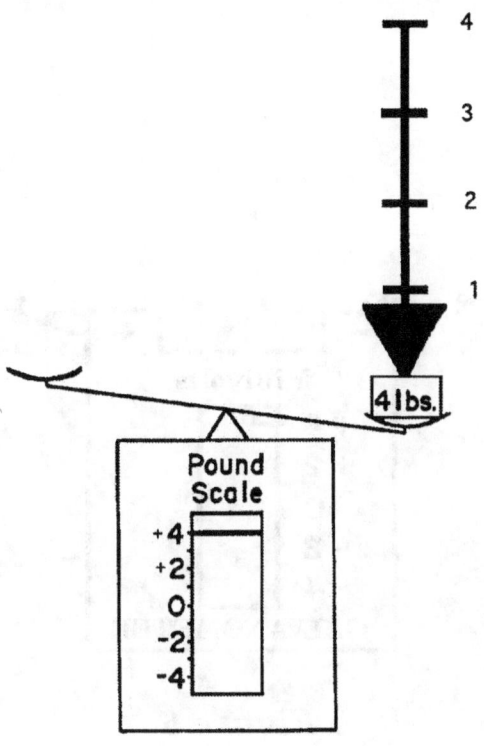

Figure 5

For the pound scale and weight, let's substitute a galvanometer and an electrical force. An arrow marked off in four equal units represents an electrical force of 4 millivolts (mV) pointing straight down (Figure 6). Let's apply the electrical force of four mV, first to the positive pole, then to the negative pole, of the galvanometer. The dial is marked off in millivolts (Figures 6 & 7).

The electrical force of four mV applied to the positive pole of the galvanometer results in a reading of +4 mV. An electrical force of four mV applied to the negative pole of the galvanometer results in a reading of minus four mV.

21

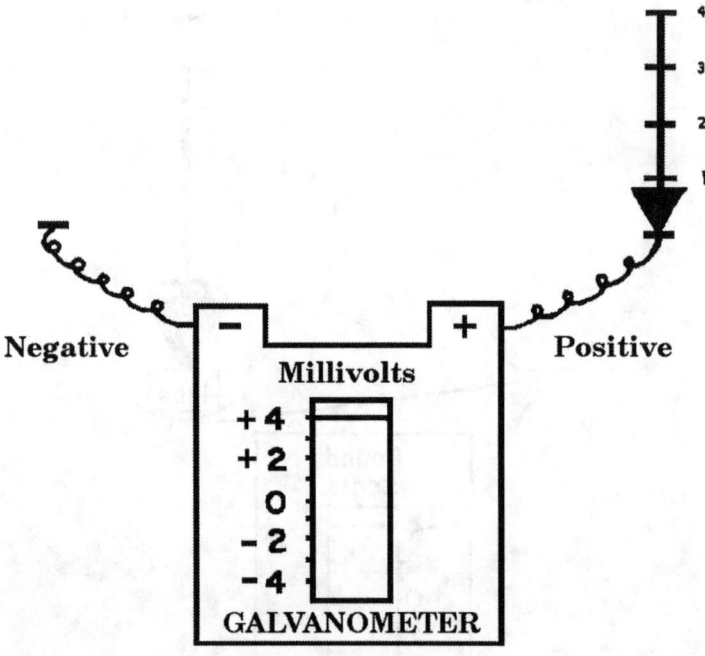

Figure 6

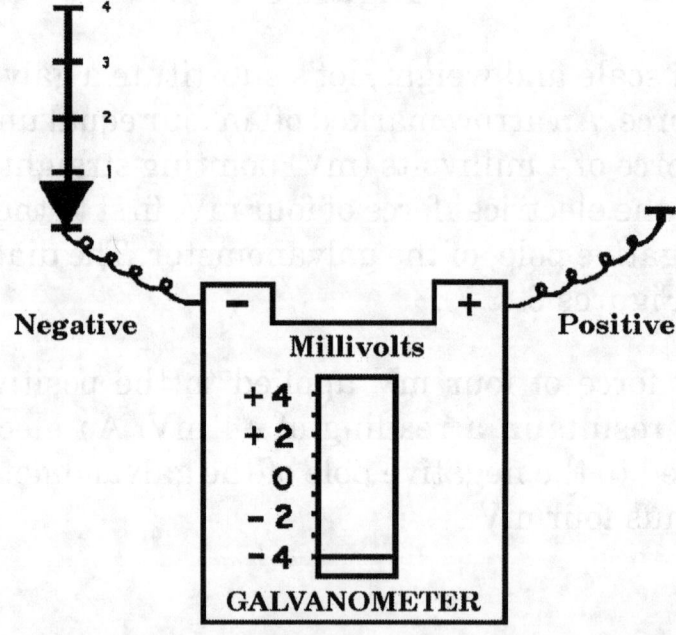

Figure 7

The galvanometer used in the electrocardiographic recorder is designed so that when an electrical force is moving toward its positive pole, an upright or positive deflection will be transcribed. When an electrical force is moving toward its negative pole, a downward deflection, a negative wave, will be recorded. These two facts, representing the basis of spatial analysis, must be memorized.

2. Cardiac Vectors

Graphs used in clinical medicine are two-dimensional documents plotting one or more patient parameters, such as, blood pressure, respirations, pulse rate and temperature, against time. The electrocardiogram, which is recorded on graph paper, appears to be two dimensional, i.e., plotting the amplitude of a deflection against time. However, the stylus of the EKG machine inscribes both upward and downward deflections, thereby adding a third dimension to the tracing, namely, direction. Direction in space is made up of three dimensions or axes (3D): left-right, superior-inferior, and anterior-posterior.

The direction of any cardiac force can be determined by noting whether the deflection is positive or negative in the lead being analyzed. These forces have magnitude and direction and are, therefore, vectors.

A positive deflection in any lead is the result of forces moving toward the positive electrode of that lead. A negative deflection in any lead is the result of forces moving toward the negative electrode of that lead. These facts are fundamental to the understanding of spatial analysis and must be memorized (Figure 8).

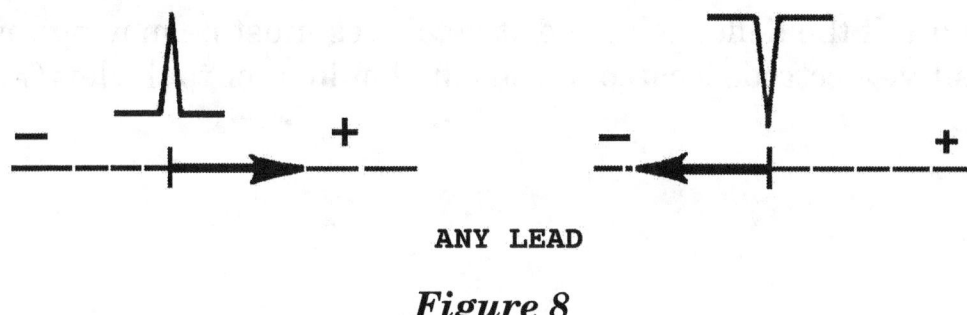

ANY LEAD

Figure 8

The right arm, left arm, and left leg play an active role in the recording of an electrocardiogram. The wires coming from these extremities are connected directly to the EKG machine. The right leg is attached to a ground terminal for electrical safety of the patient and technician. Regardless of where on the extremity an electrode is attached, the EKG machine records the tracing as if the electrode were located at the root of that extremity. The root of the left arm is the left shoulder. The root of the right arm is the right shoulder. The root of the left leg is the symphysis pubis. These are referred to as the "effective" locations of the electrodes.

B. Bipolar Limb Leads

1. Electrode Placement

a. Lead I is a bipolar limb lead with its positive electrode located on the left arm at the shoulder (LA) and negative electrode on the right arm at the shoulder (RA). Its axis is an imaginary line connecting the left and right shoulder electrodes giving it a left to right orientation. A schematic representation of Lead I is shown below. (Figure 9)

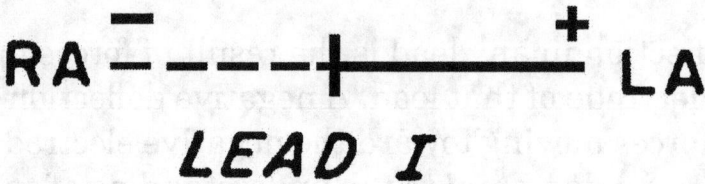

Figure 9

In Lead I, if the deflection is positive, forces must be moving toward its positive electrode located at the left shoulder, or to the left (Figure 10).

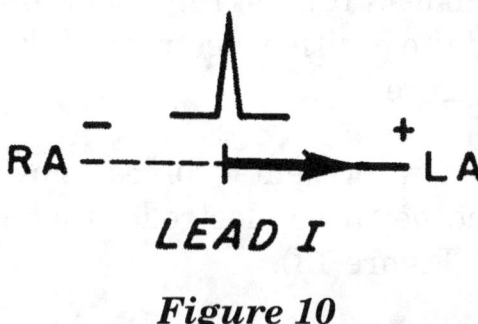

LEAD I

Figure 10

In Lead I, if the deflection is negative, forces must be moving away from its positive electrode toward its negative electrode located at the right shoulder, or to the right (Figure 11).

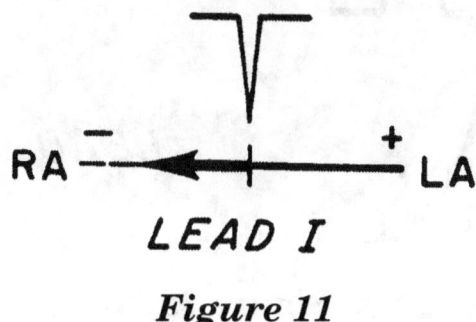

LEAD I

Figure 11

b. Lead II is a bipolar lead with its positive electrode located on the left leg. The effective location of the positive electrode is the symphysis pubis (LL). The negative electrode is located on the right arm. The effective location is the right shoulder (RA) Figure 12).

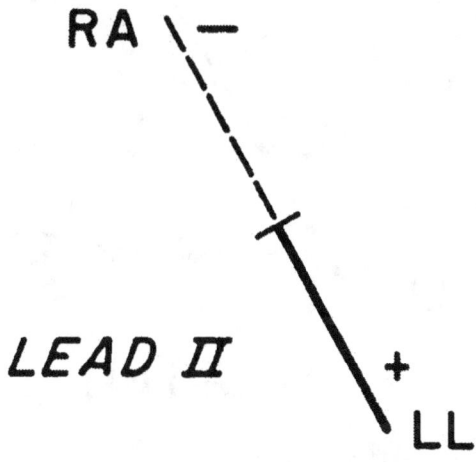

LEAD II

Figure 12

A positive deflection in lead II must represent forces moving leftward and inferiorly toward the positive electrode of this lead on the left leg (symphysis pubis) (Figure 13).

A negative deflection represents forces moving rightward and superiorly toward the negative electrode, that is, toward the right arm (right shoulder) (Figure 13).

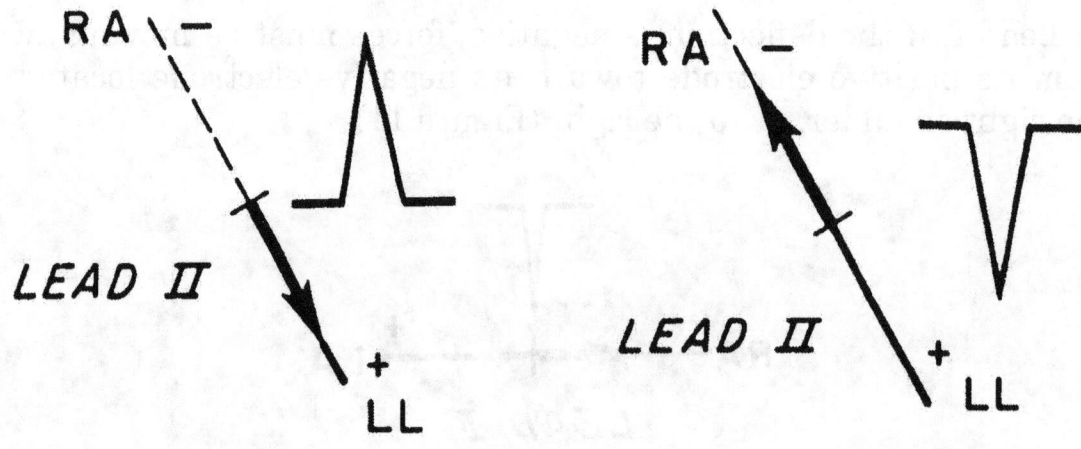

Figure 13

c. Lead III is a bipolar lead with its positive electrode located on the left leg. The effective location is the symphysis pubis (LL). The negative electrode is located on the left arm. The effective location is the left shoulder (LA) (Figure 14).

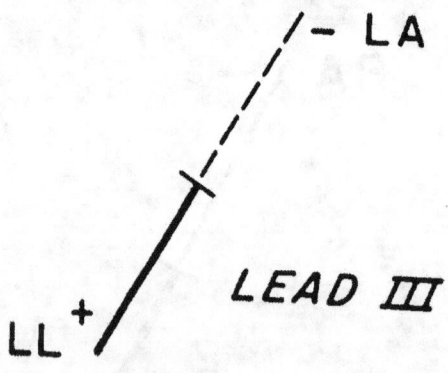

Figure 14

A positive deflection in lead III must represent forces moving rightward and inferiorly toward its positive electrode (Figure 15).

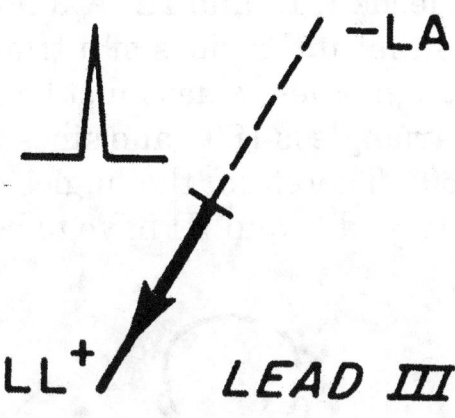

Figure 15

A negative deflection in Lead III represents forces moving leftward and superiorly toward its negative electrode (Figure 16).

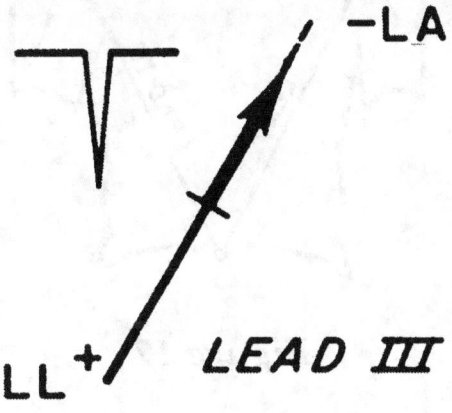

Figure 16

2. Einthoven Triangle

Leads I, II, and III are bipolar lead systems each having a positive and negative electrode. As previously stated, the effective locations of the right and left arm electrodes are at the respective shoulders, and the effective location of the left leg electrode is at the symphysis pubis. Imaginary lines joining these electrode locations, at the shoulders and the symphysis pubis, result in an electrical equilateral triangle commonly known as the Einthoven Triangle. The dot in the very center of the triangle is the "e" point, or electrical zero. The three sides of the triangle are equal in length and are represented by the axes of extremity leads I, II, and III. A fundamental theorem of plane geometry states that if the sides of a triangle are equal, then the three angles created by these sides must be also equal. Since the sum of the angles of a triangle is 180°, and since the angles are equal, each angle must be 60°. Therefore, the angles between the axes of leads I and II, I and III, and II and III have to be 60° (Figure 17).

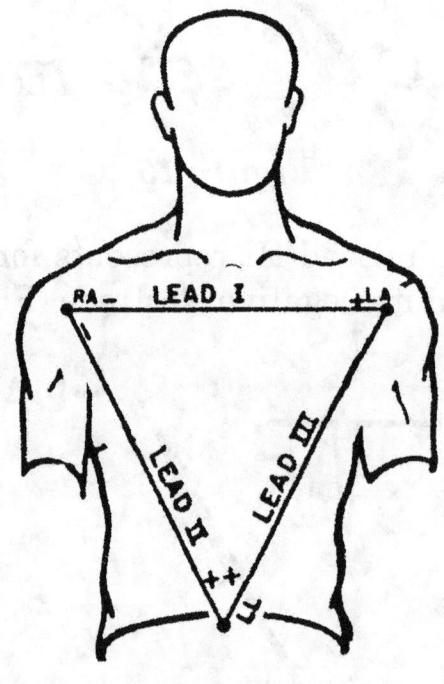

Figure 17

3. Triaxial Reference Figure

The three axes shown in Figure 18 may be moved in such a way so that they pass through a common point, the electrical center or "e" point of the heart. The next illustration demonstrates how this is done.

The geometric figure formed by the axes of the three standard limb leads is shown below on the left. The sides are equal and the figure is known as an "equilateral triangle". The triaxial reference figure is formed by shifting the three axes, without changing their direction, so that all three pass through a common point, the "e" point of the heart. This is illustrated on the right. When re-drawn in this manner, the limb lead axes form a figure consisting of three axes separated by 60°

from one another. They constitute the Triaxial Reference Figure of Bayley. The angle in degrees and the polarity of each lead (location of the positive and negative electrodes) have been added to the Triaxial Reference Figure.

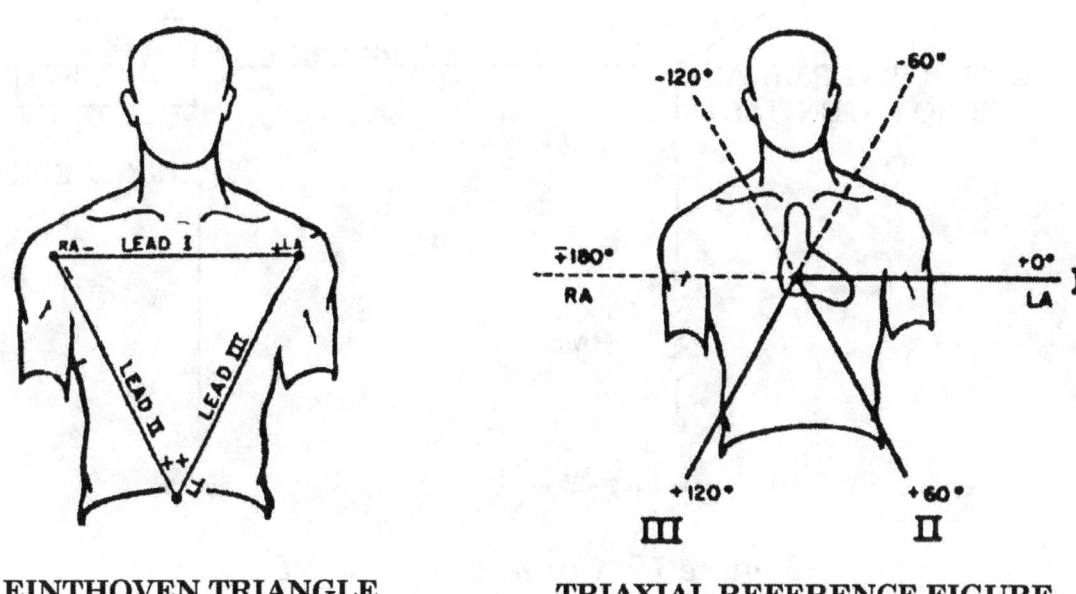

EINTHOVEN TRIANGLE   **TRIAXIAL REFERENCE FIGURE**

Figure 18

C. Unipolar Limb Leads

The unipolar extremity leads are recorded with the positive electrodes located on their respective extremities, the right arm (right shoulder), left arm (left shoulder), and left leg (symphysis pubis). The negative electrode of each unipolar limb lead is a central terminal at zero volts, devised by Wilson. The lead wires from each of the three extremities are connected through equal resistances to a central terminal. The voltage of the central terminal remains at zero throughout the cardiac cycle. Pairing the central terminal with an exploring limb electrode gives rise to deflections that are too small and difficult to read. Because of this, augmented unipolar limb leads aVR, aVL, and aVF were developed. This was done by removing the connection between the central terminal and the extremity whose deflections were being measured. This circuit change increases the amplitude of the deflections by 50% without altering their form. This is illustrated in Figures 19 and 20.

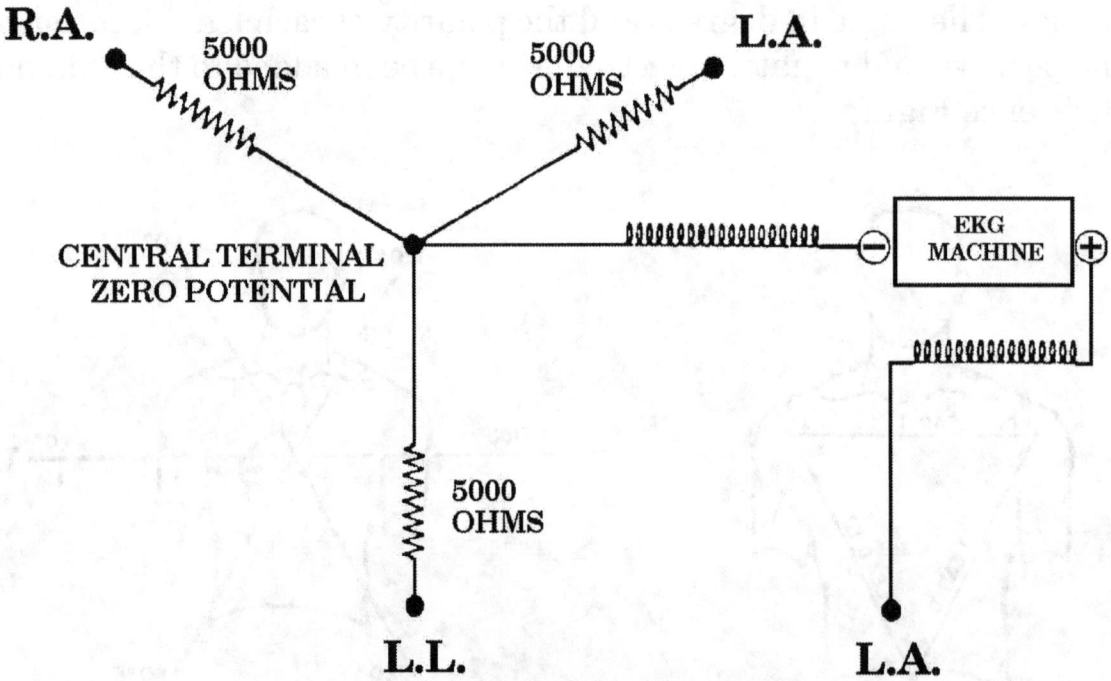

Figure 19: Connection for VL

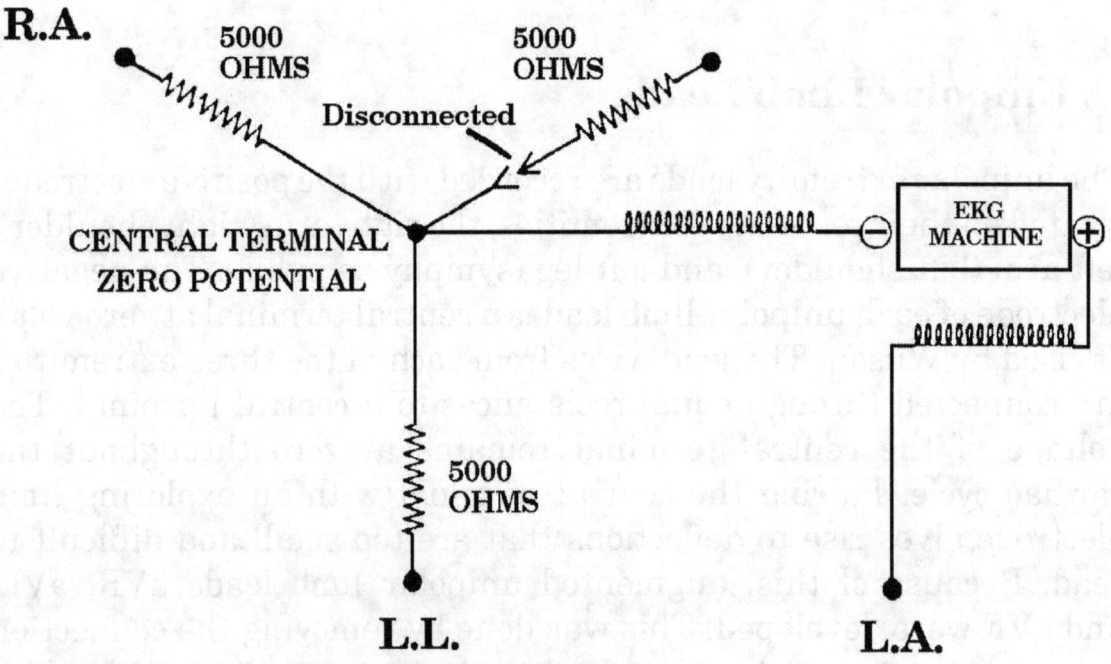

Figure 20: Connection for aVL

The unipolar limb leads, aVR, aVL, and aVF, are recorded with their positive electrodes located on their respective extremities: aVR, the right arm at the right shoulder (RA), aVL, the left arm at the left shoulder (LA), and aVF, the left leg at the symphysis pubis (LL). The negative electrode is a central terminal assumed to be located at the electrical center of the heart, the "e" point. The effective spatial location of the negative electrode is 180° away from the positive electrode (Figure 21).

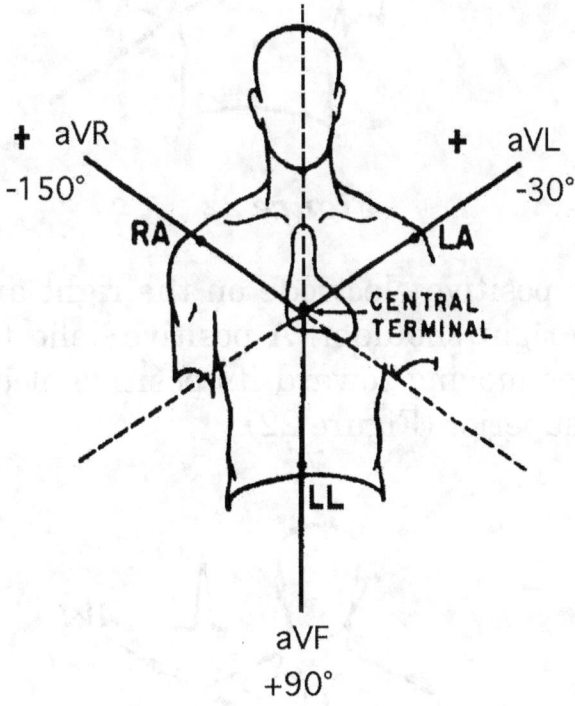

Figure 21

1. Electrode Placement

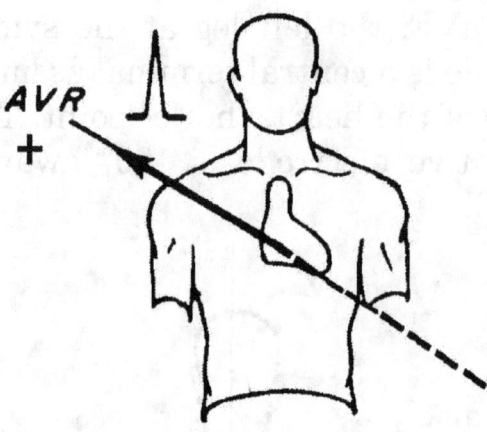

Figure 22

a. aVR has its positive electrode on the right arm. The effective location is the right shoulder. A positive deflection in aVR must represent forces moving toward its positive electrode, that is, to the right and superior (Figure 22).

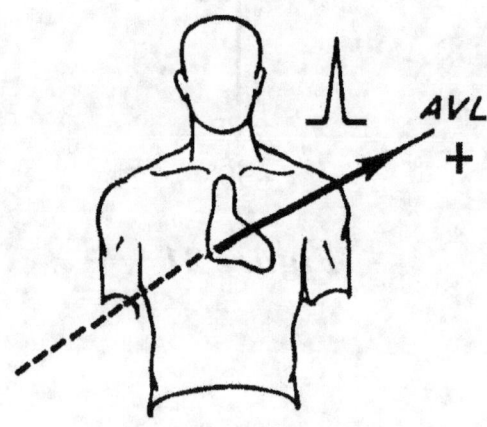

Figure 23

b. aVL has its positive electrode on the left arm. The effective location is the left shoulder. A positive deflection in aVL means forces are moving toward its positive electrode, that is, left and superior (Figure 23).

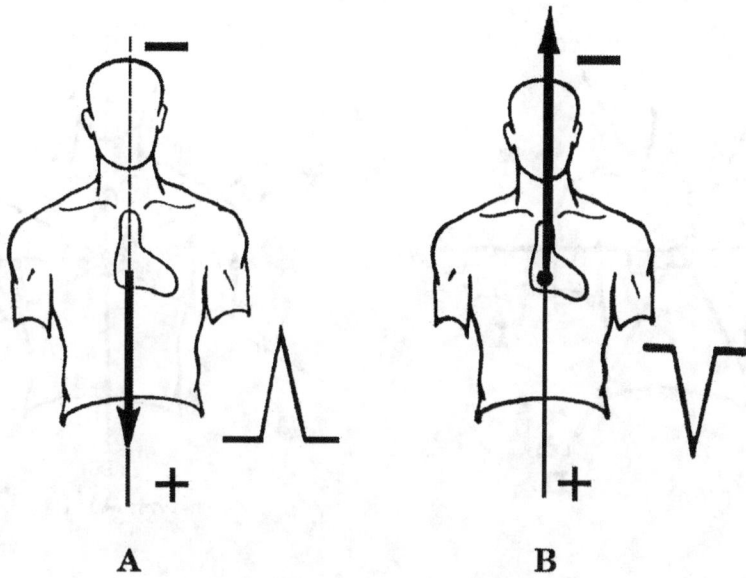

A   B

Figure 24: Lead aVF

c. aVF has its positive electrode on the left leg. The effective location is the symphysis pubis. An upright or positive deflection in Lead aVF represents forces moving toward its positive electrode, the symphysis pubis, and therefore, inferiorly (Figure 24A). A negative deflection in Lead aVF represents forces moving away from the positive electrode and toward the negative electrode, therefore, superiorly (Figure 24B).

2. Hexaxial Reference Figure

As previously stated, when the axes of the Standard Bipolar Limb Leads I, II, and III are drawn through a common point, the electrical center of the heart, the Triaxial Reference Figure of Bayley is produced (Figure 25). When the axes of unipolar limb leads aVR, aVL, and aVF are superimposed on the triaxial reference figure, the hexaxial reference figure, consisting of six axes separated by 30°, is created (Figure 26). By convention, the angle of each axis is arranged as shown in Figure 26.

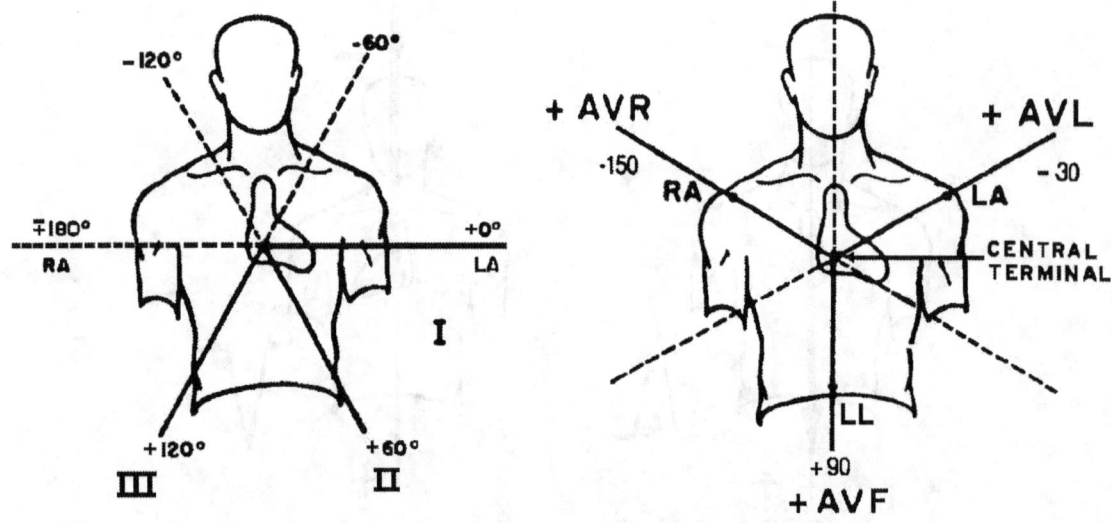

Figure 25

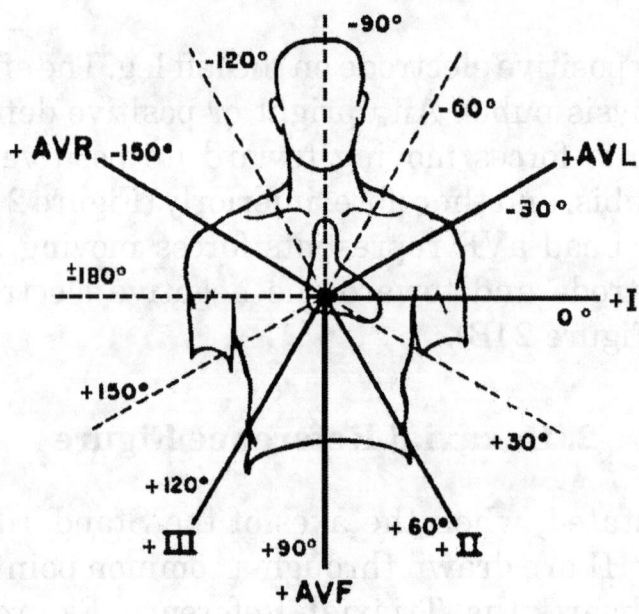

Figure 26

3. Frontal Plane

The axes of bipolar limb leads I, II, III, and unipolar limb leads aVR, aVL, and aVF lie in the frontal plane. This plane divides the body into a front half and a back half. The left—right axis of Lead I and the superior—inferior axis of Lead aVF represent the major coordinates of the frontal plane (Figure 27).

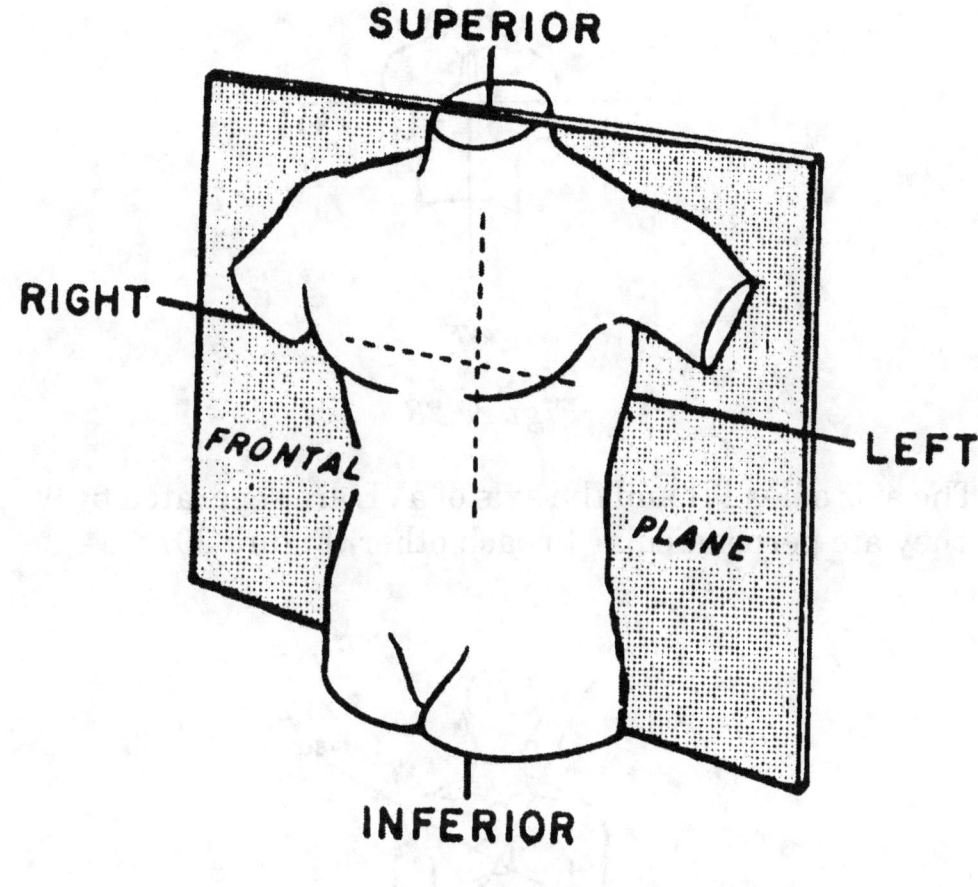

Figure 27

4. Perpendiculars

a. The axis of lead I and the axis of lead aVF are separated by 90°, that is, they are perpendicular to each other (Figure 28).

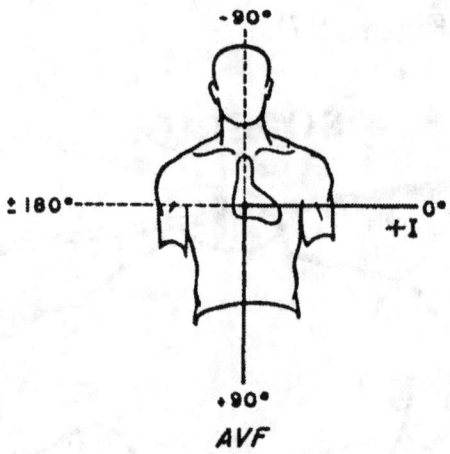

Figure 28

b. The axis of lead II and the axis of aVL are separated by 90°, that is, they are perpendicular to each other (Figure 29).

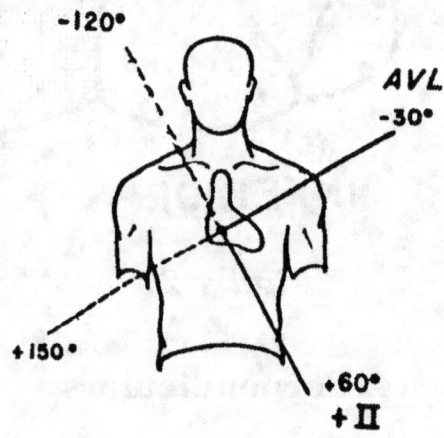

Figure 29

c. The axis of lead III and the axis of aVR are separated by 90°, that is, they are perpendicular to each other (Figure 30).

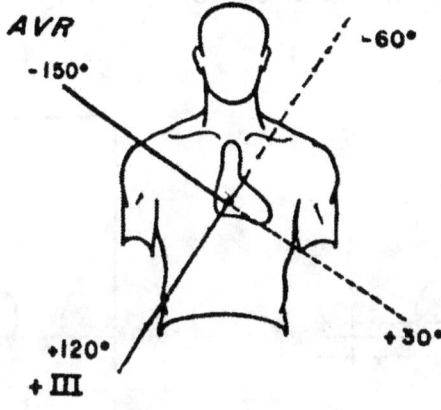

Figure 30

5. Axes and Fields of Bipolar Leads

The axis of Lead I may be divided into a positive and negative half by a perpendicular line through its center point as illustrated in Figure 31.

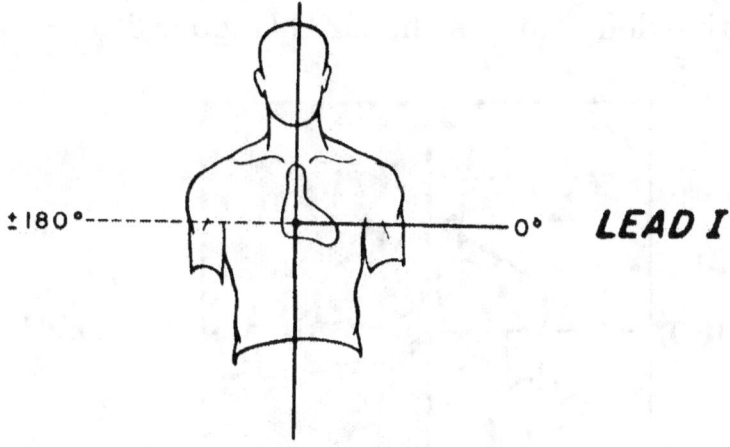

Figure 31

Since the positive electrode of Lead I is located on the left shoulder, the positive half of the axis must lie to the left of its perpendicular line (the axis of Lead aVF). The positive end of the axis of Lead I is set arbitrarily at 0° and the negative portion is located to the right at 180° (Figure 32A). The perpendicular line through the center of the axis of Lead I divides the body into a positive and negative field. The two quadrants making up the positive field of Lead I are stippled, and the two quadrants comprising the negative field are white (Figure 32B).

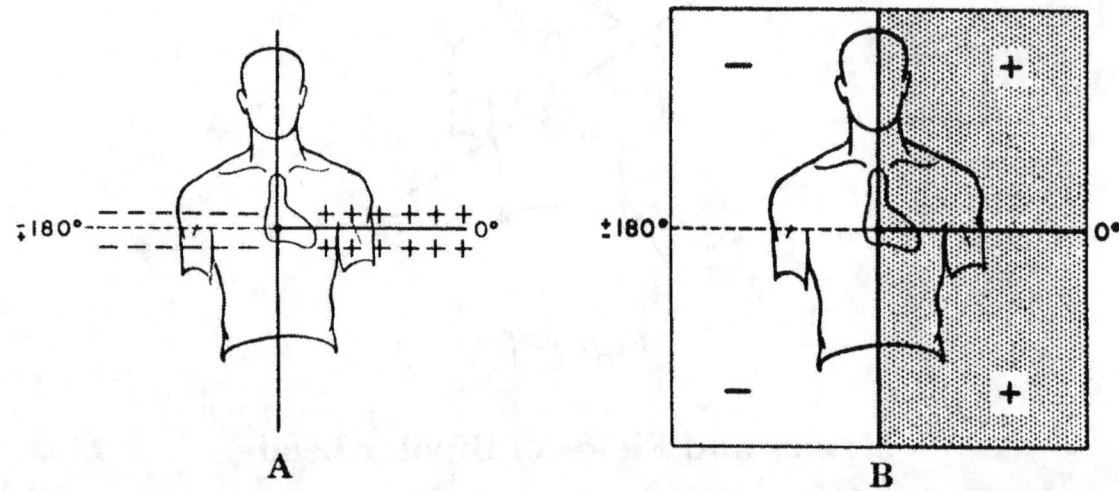

Figure 32: A & B

A QRS vector located within the positive field of lead I must represent forces moving toward the positive electrode of lead I, and will give rise to a positive deflection on the EKG (Figure 33).

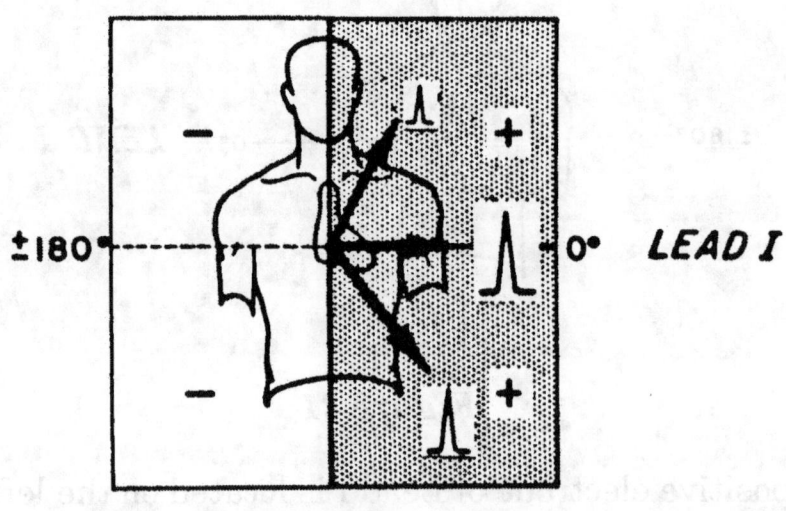

Figure 33

A QRS force or vector located within the negative field of lead I must represent forces moving toward the negative electrode of lead I, and will give rise to a negative deflection on the EKG (Figure 34).

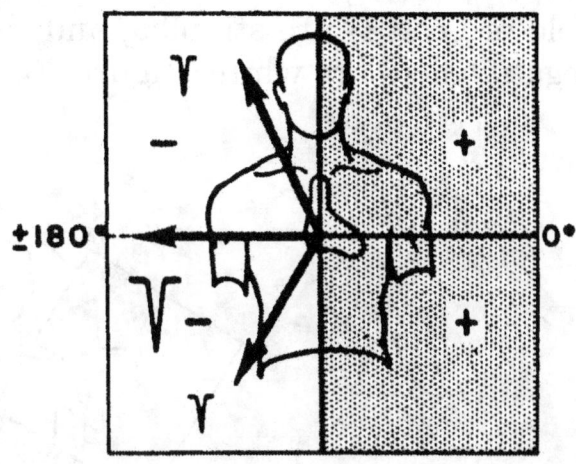

Figure 34

The axis of Lead II may be divided into a positive and negative half by a perpendicular line through its center point as shown in Figure 35.

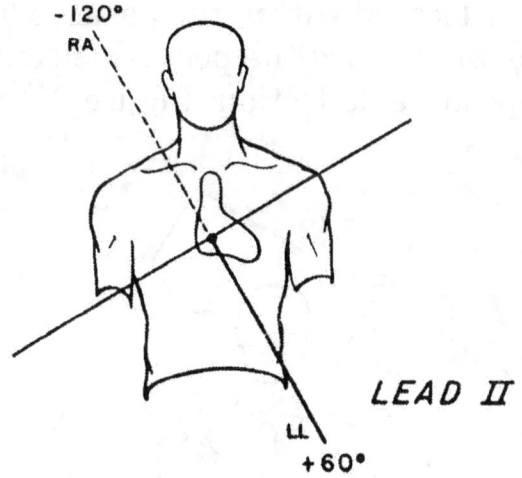

Figure 35

Since the positive electrode of Lead II is located on the left leg (symphysis pubis), the positive portion of the axis of Lead II must lie at the left and inferior side of its perpendicular (the axis of Lead aVL). The positive end of the axis of Lead II is 60° away from the positive end of Lead I, and, therefore, at +60°. The negative portion is located right and superior at −120° (Figure 35). The perpendicular line through the center of the axis of Lead II divides the body into a positive and negative field (Figure 36A). The two quadrants making up the positive field of Lead II are stippled, and the two quadrants comprising the negative field are white (Figure 36B).

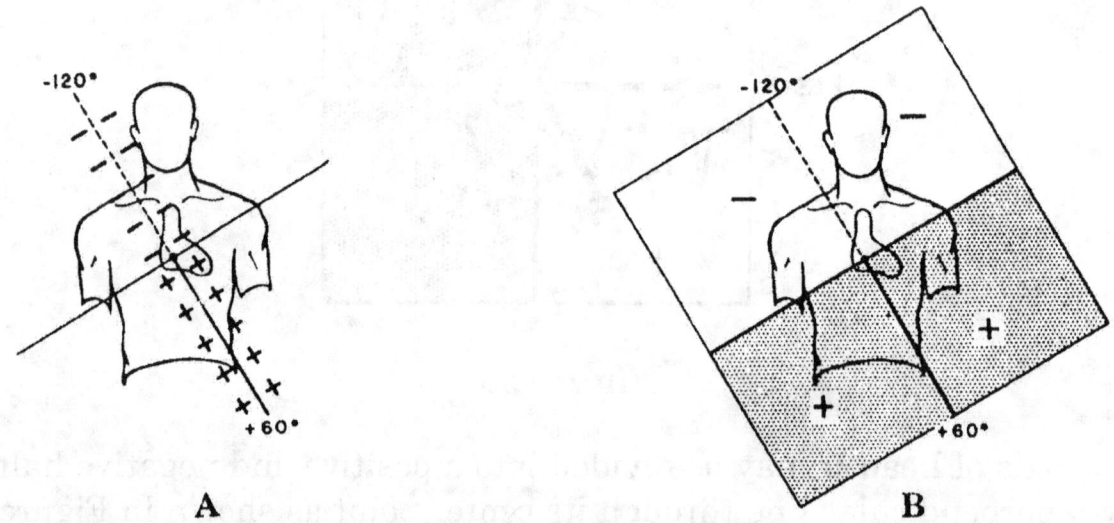

Figure 36: A, B

A QRS force or vector located within the positive field of lead II must represent forces moving toward the positive electrode of lead II, and must give rise to a positive deflection (Figure 37).

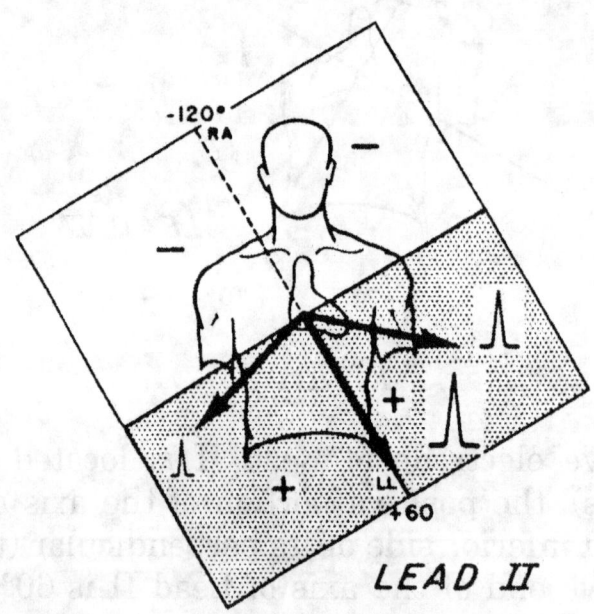

Figure 37

A QRS force or vector located within the negative field of lead II must represent forces moving toward the negative electrode of lead II, and must give rise to a negative deflection Figure 38).

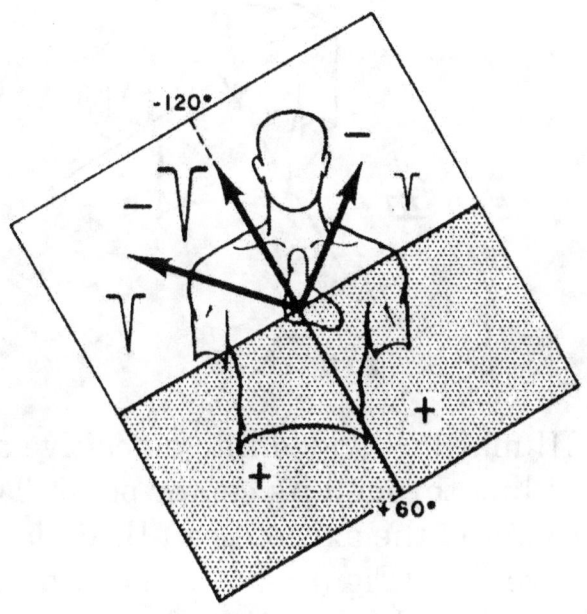

Figure 38

Since the positive electrode of Lead III is located on the left leg (symphysis pubis), the positive portion of the axis of Lead III must lie on the right and inferior side of its perpendicular line (the axis of Lead aVR). The positive end of the axis of Lead III is 60° away from the positive end of Lead II at +120°. The negative portion is located to the left and superior at –60° (Figure 39).

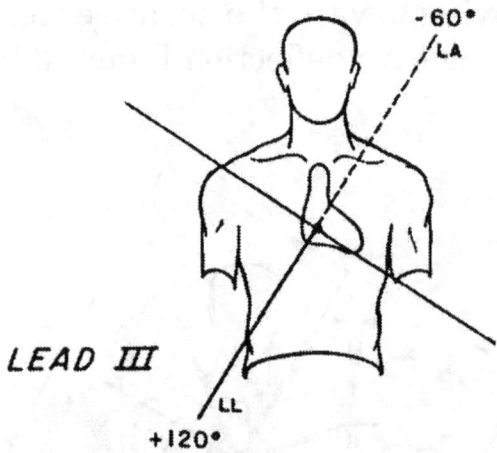

Figure 39

The axis of Lead III may be divided into a positive and negative half by a perpendicular line through its center point. The perpendicular line through the center of the axis of Lead III divides the body into a positive and negative field (Figure 40A). The two quadrants making up the positive field of Lead III are stippled, and the two quadrants comprising the negative field are white (Figure 40B).

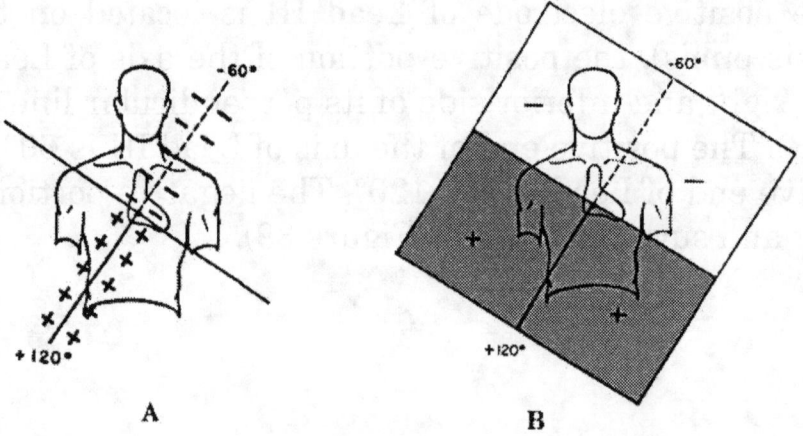

A   B

Figure 40: A, B

A QRS force or vector located within the positive field of lead III must represent forces moving toward the positive electrode of lead III, and must give rise to a positive deflection Figure 41).

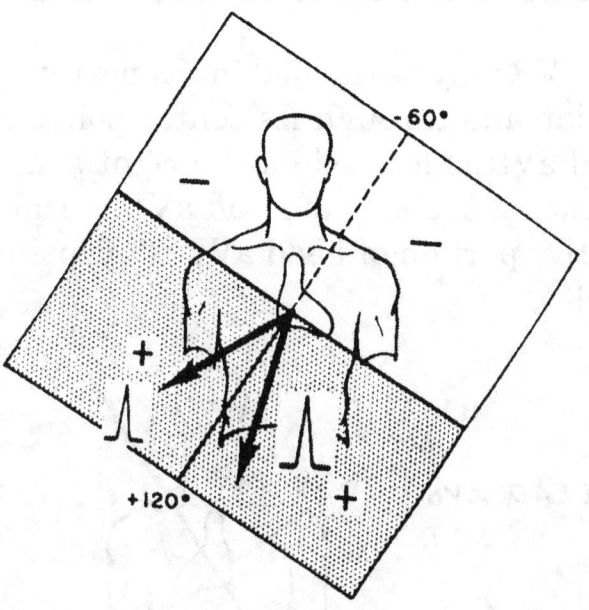

Figure 41

A QRS force or vector located within the negative field of lead III must represent forces moving toward the negative electrode of lead III, and must give rise to a negative deflection (Figure 42).

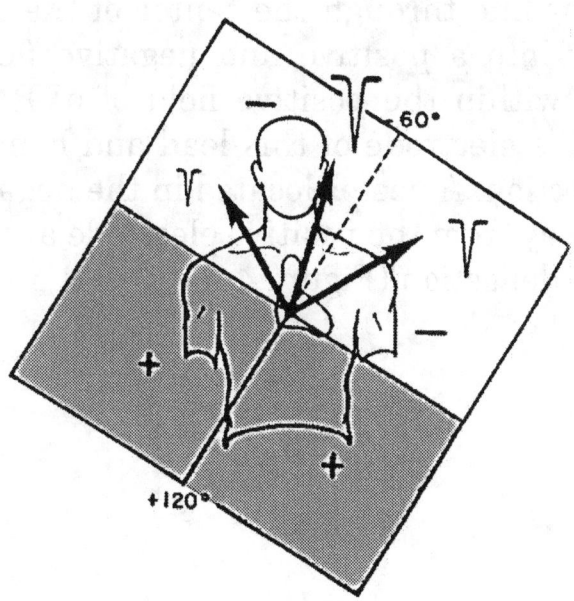

Figure 42

6. Axes and Fields of Unipolar Leads

The axis of Lead aVR may be divided into a positive and negative half by a perpendicular line through its center point. Since the positive electrode of Lead aVR is located on the right arm (right shoulder), the positive portion of the axis of Lead aVR is right and superior at −150°. The negative portion of Lead aVR is opposite, left and inferior at +30° (Figure 43).

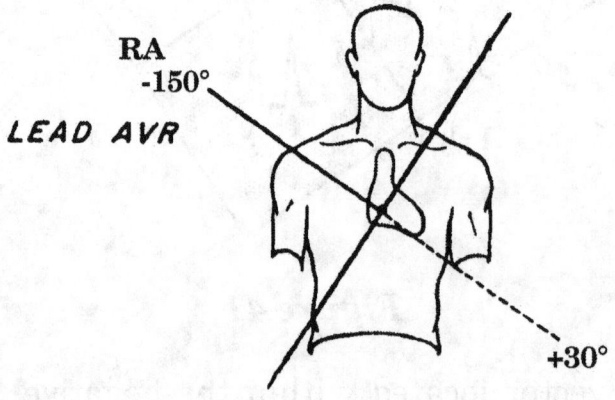

Figure 43

The perpendicular line through the center of the axis of Lead aVR divides the body into a positive and negative field (Figure 44A). A vector located within the positive field of aVR must be moving toward the positive electrode of this lead and hence must give rise to a positive deflection. A vector located in the negative field of Lead aVR is moving away from the positive electrode and hence must give rise to a negative deflection (Figure 44B.

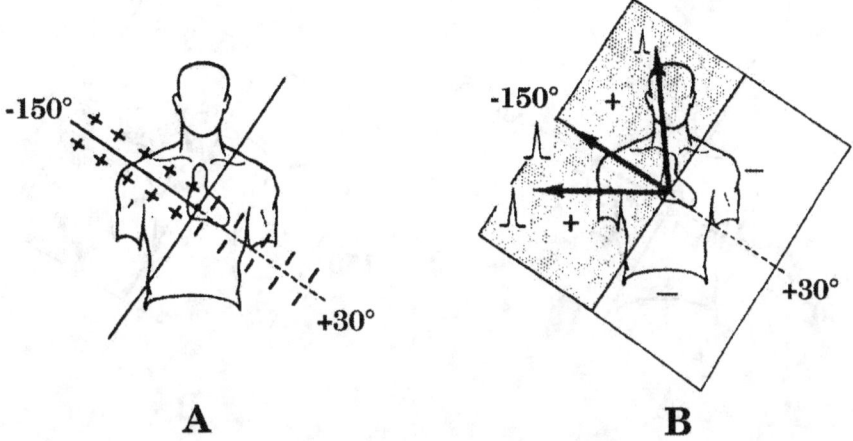

A   **B**

Figure 44 A, B

The axis of Lead aVL may be divided into a positive and negative half by a perpendicular line through its center point (A). Since the positive electrode of Lead aVL is located on the left arm at the left shoulder, the positive portion of the axis of Lead aVL is left and superior at −30° (B). The negative portion of Lead aVL is opposite, right and inferior, at +150° (Figure 45).

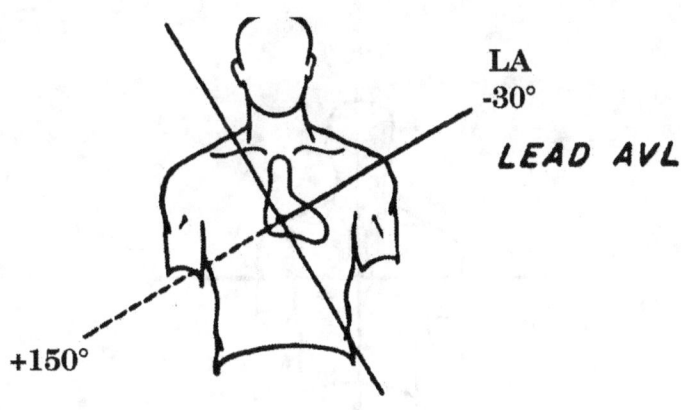

Figure 45

The perpendicular line through the center of the axis of Lead aVL also divides the body into a positive and negative field (Figure 46A)). A vector lying within the positive field of aVL must be moving toward the positive electrode of this lead and, therefore, must give rise to a positive deflection (Figure 46B)). A vector located within the negative field of Lead aVL is moving away from the positive electrode and hence will give rise to a negative deflection (not shown).

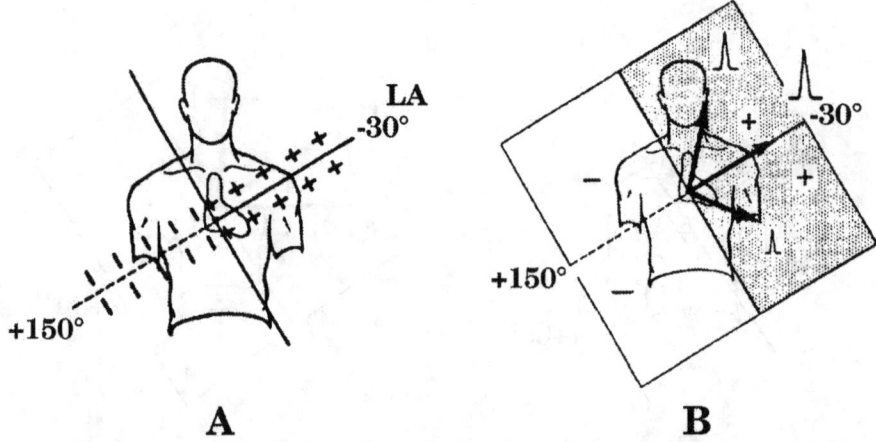

A **B**

Figure 46 A, B

The axis of Lead aVF may be divided into a positive and negative half by a perpendicular line through its center point. Since the positive electrode of Lead aVF is located on the left leg (symphysis pubis), the positive portion of the axis of Lead aVF is inferior at +90°. The negative electrode of Lead aVF is a central terminal whose spatial location is 180° away from its positive electrode at +90° (Figure 47).

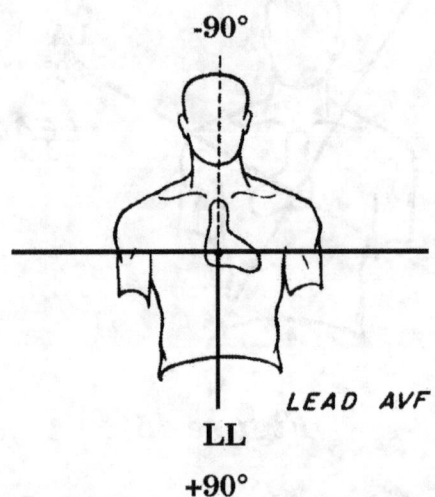

Figure 47

The perpendicular line through the center of the axis of Lead aVF also divides the body into a positive and negative field (Figure 48A). A vector located within the positive field of aVF must be moving inferiorly toward the positive electrode of this lead and, therefore, must give rise to a positive deflection (Figure 48B). A vector located within the negative field of Lead aVF is moving superiorly away from the positive electrode and hence will give rise to a negative deflection (not shown).

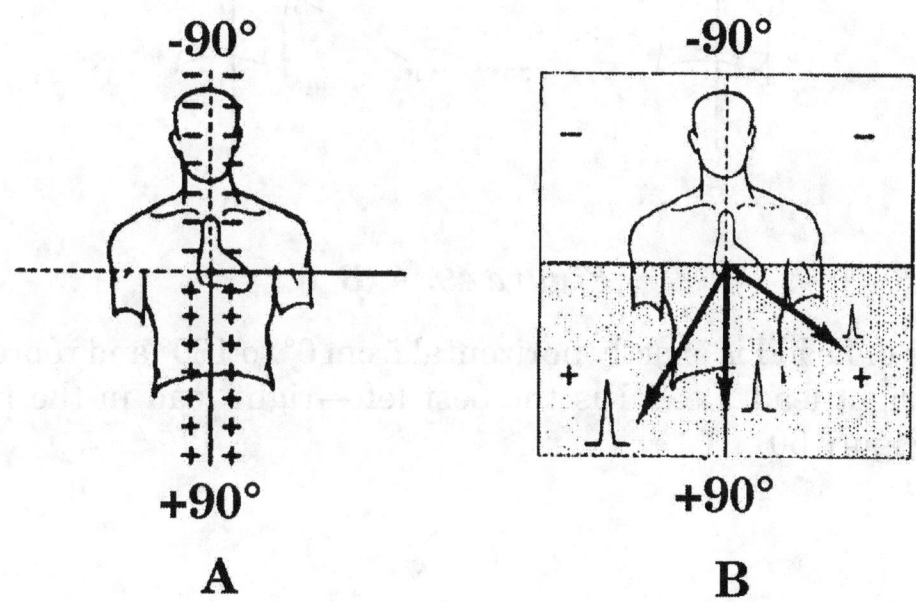

Figure 48 A, B

7. Best Left—Right Limb Lead

The location of the positive electrode of Lead aVR is on the right shoulder. The positive end of the axis of Lead aVR is oriented right and superior at −150°. This axis is tilted 30° from the horizontal left—right axis and 60° from the vertical. Lead aVR, therefore, is primarily a left—right lead with a superior—inferior tilt and not the best left—right lead (Figure 49A).

The location of the positive electrode of Lead aVL is on the left shoulder. The positive end of the axis of Lead aVL is oriented left and superior at −30°. This axis is tilted 30° from the left—right axis and 60° from the superior—inferior axis. Lead aVL, therefore, is primarily a left—right lead with a superior—inferior tilt and not the best left—right lead (Figure 49B).

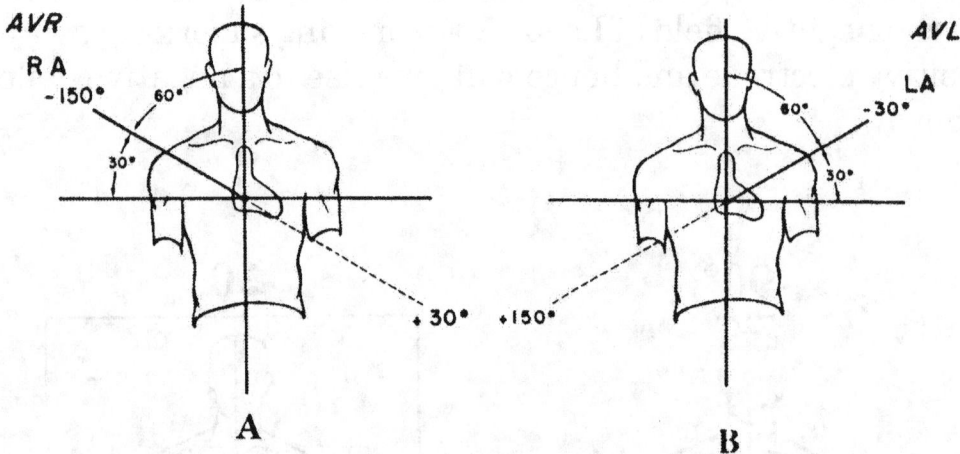

Figure 49: A, B

The axis of Lead I is exactly horizontal from 0° to 180° and represents a left—right lead. Lead I is the best left—right lead in the frontal plane (Figure 50).

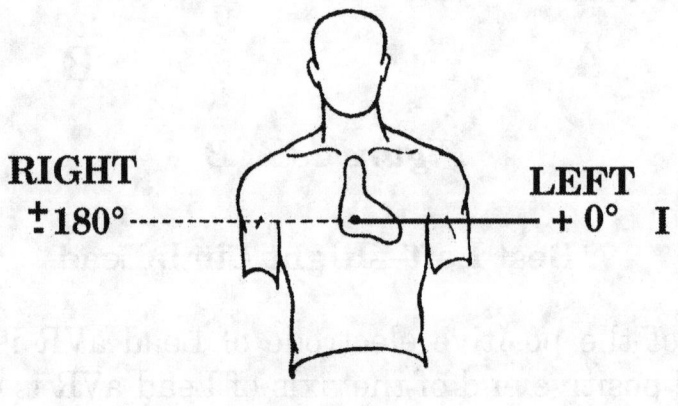

Figure 50

8. Best Superior—Inferior Limb Lead

The positive end of the axis of Lead II is located at +60° and the negative end at −120°. Lead II is primarily a superior—inferior lead with a tilt. It is not, therefore, the best superior—inferior lead (Figure 51).

The positive end of the axis of Lead III is located at +120° and the negative end at −60°. Lead III is primarily a superior—inferior lead with a tilt. It is not, therefore, the best superior—inferior lead (Figure 51B).

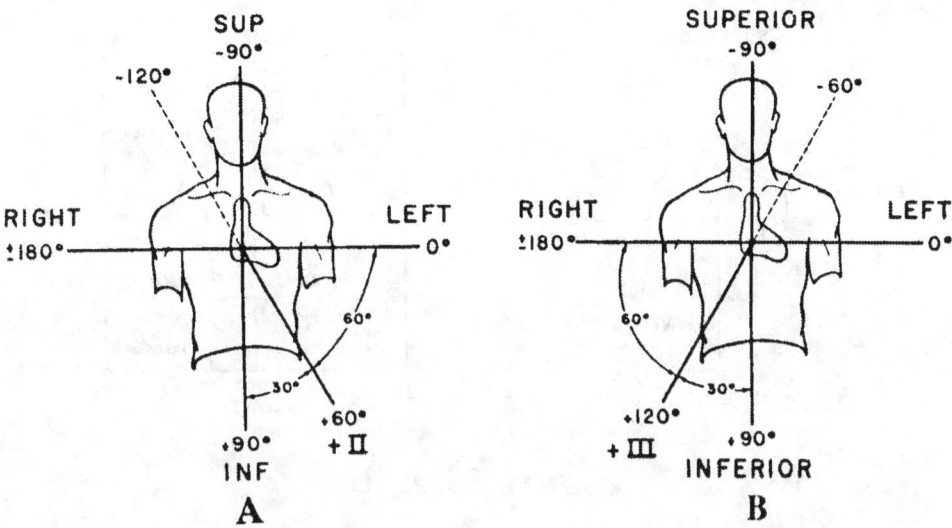

Figure 51: A, B

Lead aVF is oriented along the +90° to −90° axis and represents a superior–inferior lead with no tilt. Lead aVF is, therefore, the best superior–inferior lead in the frontal plane (Figure 52).

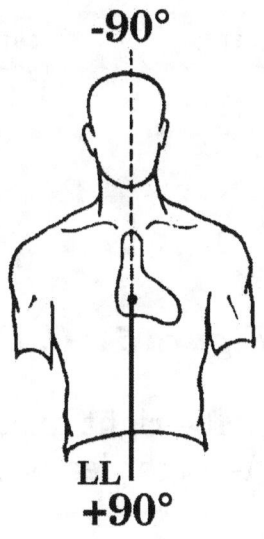

Figure 52

9. Best Frontal Plane Reference Leads

Leads I and aVF are the best frontal plane leads for determining vector data along the left—right and superior—inferior axes (Figure 53A). These axes define the frontal plane (Figure 53B).

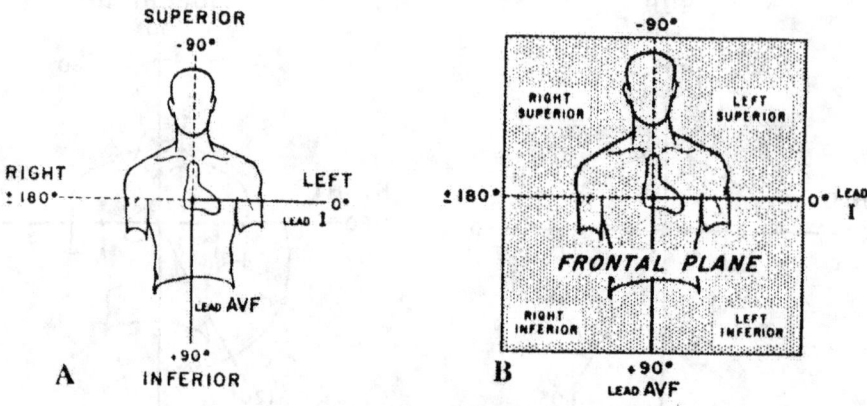

Figure 53: A, B

Leads II and III have superior—inferior axes with a left—right tilt and are not the best superior—inferior leads (Figure 54 C, D).

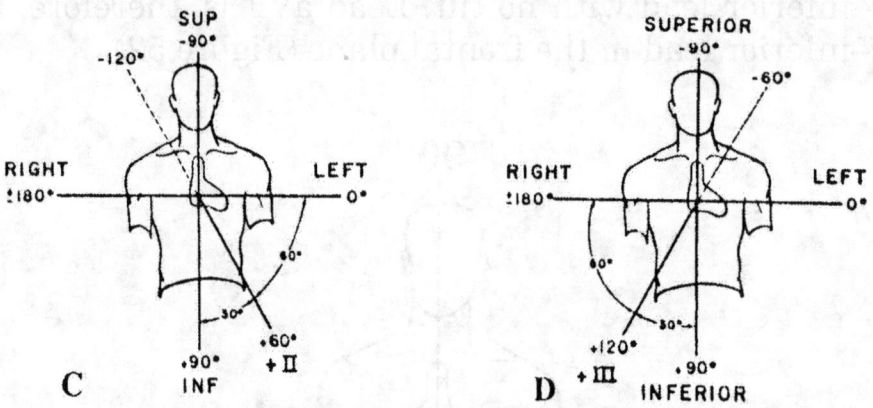

Figure 54: C, D

Leads aVR and aVL have left—right axes with a superior—inferior tilt and are not the best left—right leads (Figure 55 E, F).

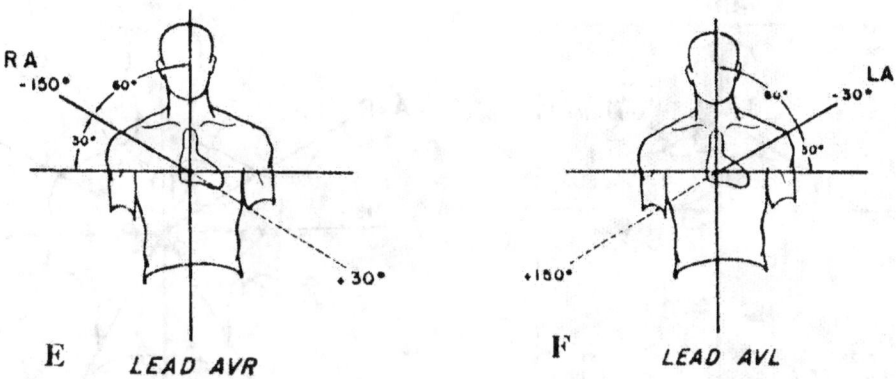

Figure 55: E, F

10. Quadrant and Perpendicular Rules

The mean frontal vector of any electrocardiographic deflection such as, a P wave, QRS complex, ST segment, or T wave, may be located using the quadrant and perpendicular rules of spatial analysis. The Quadrant Rule is used to localize the mean vector to one of the four frontal plane quadrants. This is best accomplished by examining the deflections in Lead I for left—right information and in Lead aVF for superior—inferior information. (Figure 56A). Once the deflection is localized to the appropriate quadrant, the Perpendicular Rule is employed. This rule states that the mean vector lies in the predetermined quadrant perpendicular to the axis of the lead that exhibits the equiphasic complex. (Figure 56B). A complex is equiphasic when the positive and negative **areas** are equal, that is, algebraically zero (Figure 57).

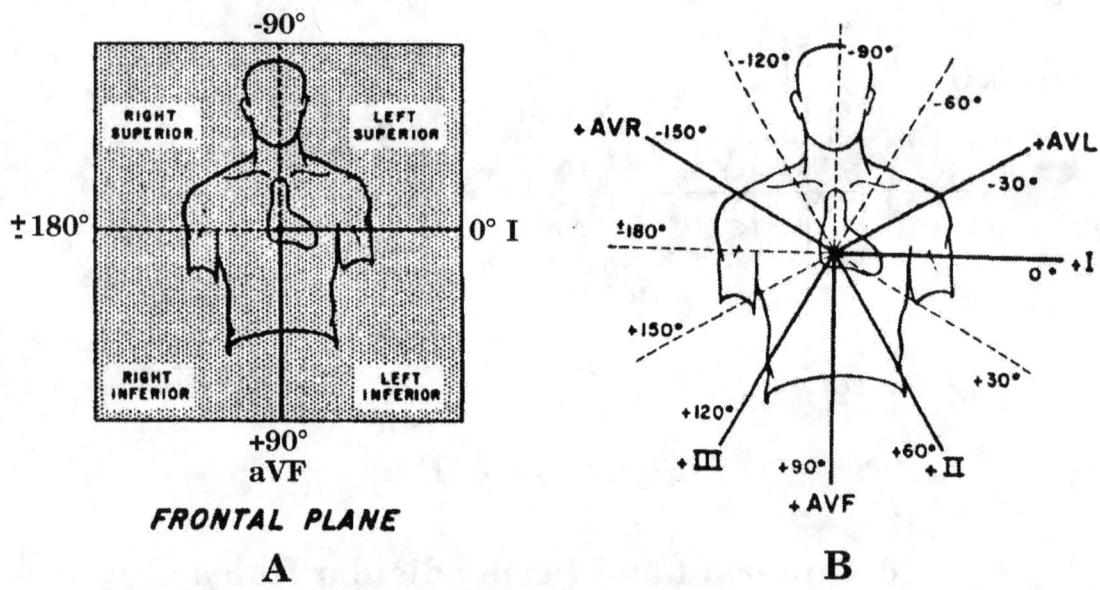

Figure 56: A, B

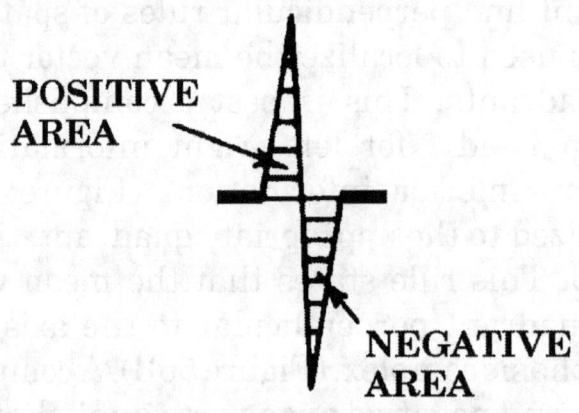

Figure 57: Equiphasic Complex

C. Mean Frontal Vectors

1. QRS Vector

Each QRS complex in the electrocardiogram represents depolarization of the ventricles. During activation of the ventricles, an infinite number of vectors is produced and conducted to all regions of the body, including the skin, and are recorded in all lead systems (Figure 58).

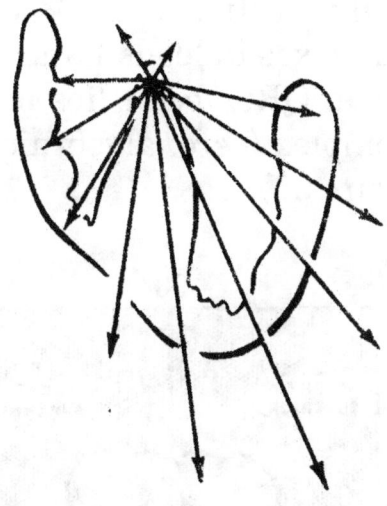

Figure 58

All these instantaneous vectors produced during one cycle of ventricular activation may be added, resulting in a single QRS vector having a specific direction in space. This single vector (thick bold arrow) is referred to as the mean QRS vector. The direction of the mean QRS vector is important in the diagnosis of cardiac abnormalities (Figure 59).

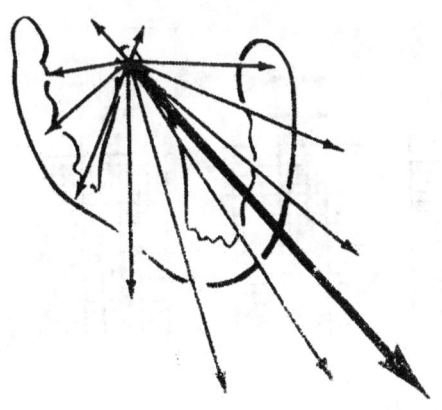

Figure 59

2. Mean Frontal QRS Vector

The mean frontal QRS vector of any electrocardiogram may be located by the Quadrant and Perpendicular Rules of Spatial Analysis. The quadrant rule localizes the mean QRS vector to one of the four frontal plane quadrants (see illustration below). This is accomplished by looking at the QRS complexes in leads I and aVF. The perpendicular rule states that the mean QRS vector lies perpendicular to the lead with the equiphasic complex (zero, algebraic or actual) in the pre—selected quadrant (Figure 60).

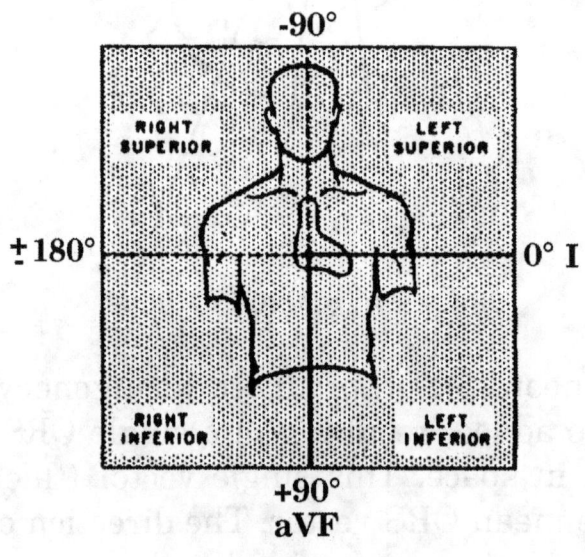

FRONTAL PLANE

Figure 60

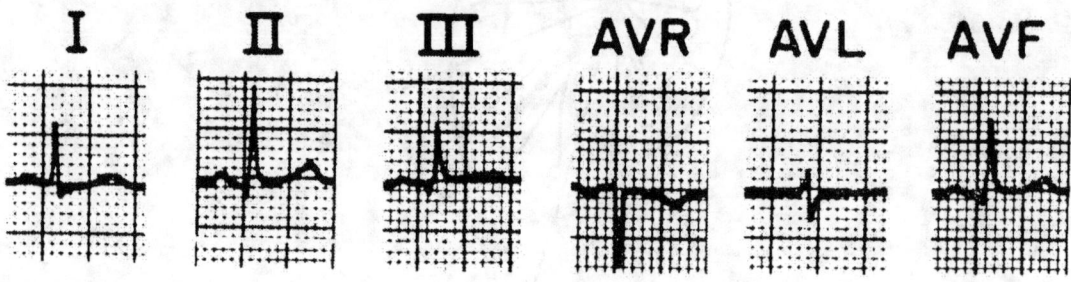

Example 1

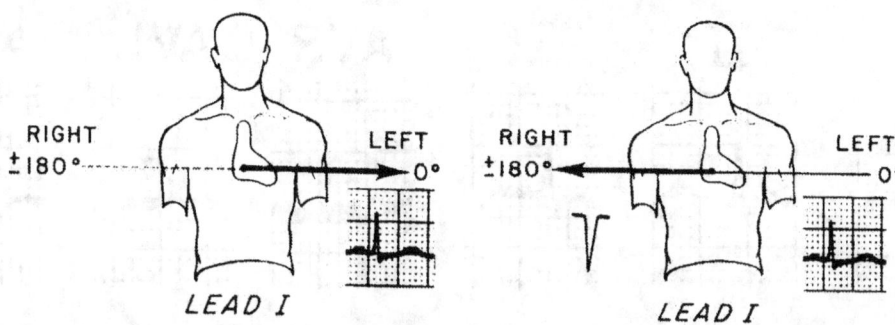

Figure 61

In the EKG illustrated in Example 1, Lead I has an R wave which indicates QRS forces are moving toward the positive electrode of Lead I, or to the left. The vector cannot be moving to the right since this would demand a negative deflection in Lead I (Figure 61).

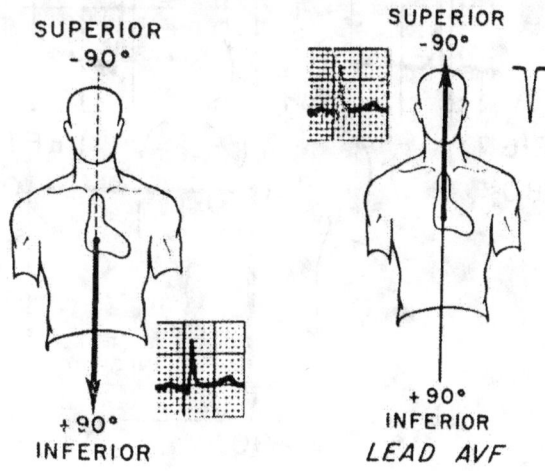

Figure 62

The predominant QRS deflection in Lead aVF is positive indicating that QRS forces are moving inferiorly toward its positive electrode at the symphysis pubis. This vector cannot be moving superiorly since this would demand a negative QRS deflection in Lead aVF (Figure 62).

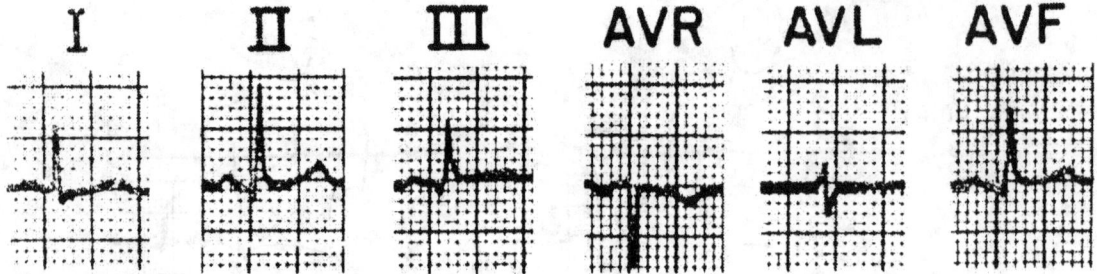

The QRS complex is positive in I and aVF indicating that the forces are directed leftward and inferiorly in the left inferior quadrant (Figure 63).

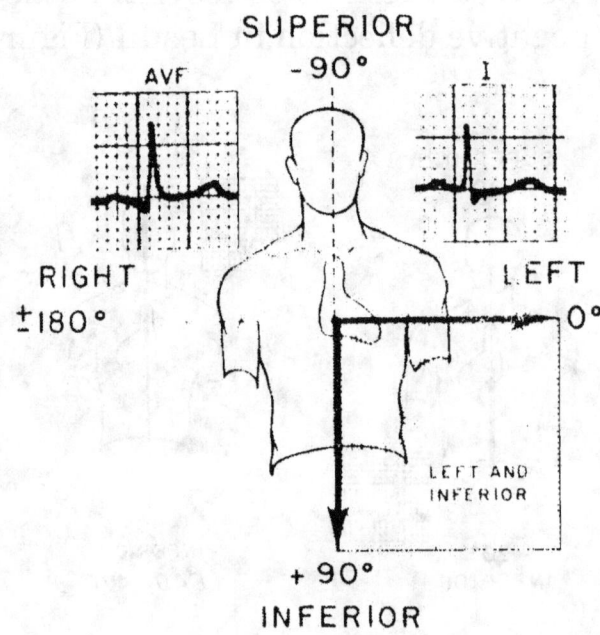

Figure 63

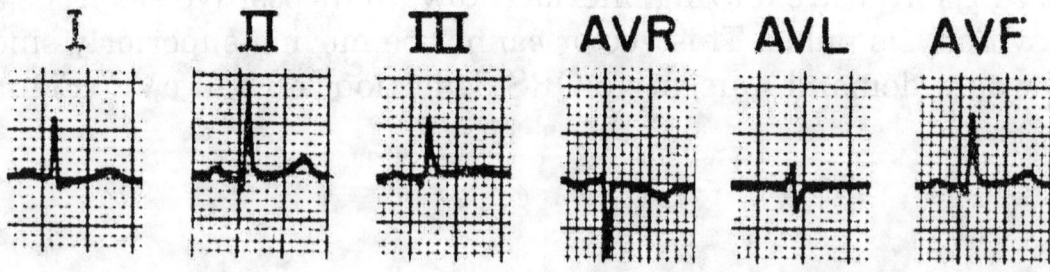

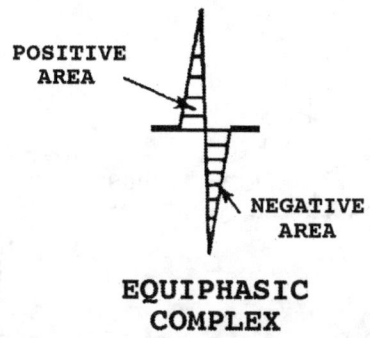

POSITIVE AREA

NEGATIVE AREA

EQUIPHASIC COMPLEX

Figure 64

Since the QRS complex in aVL is clearly equiphasic, the mean QRS vector must be perpendicular to the axis of aVL in the left inferior quadrant at +60° (Figure 65).

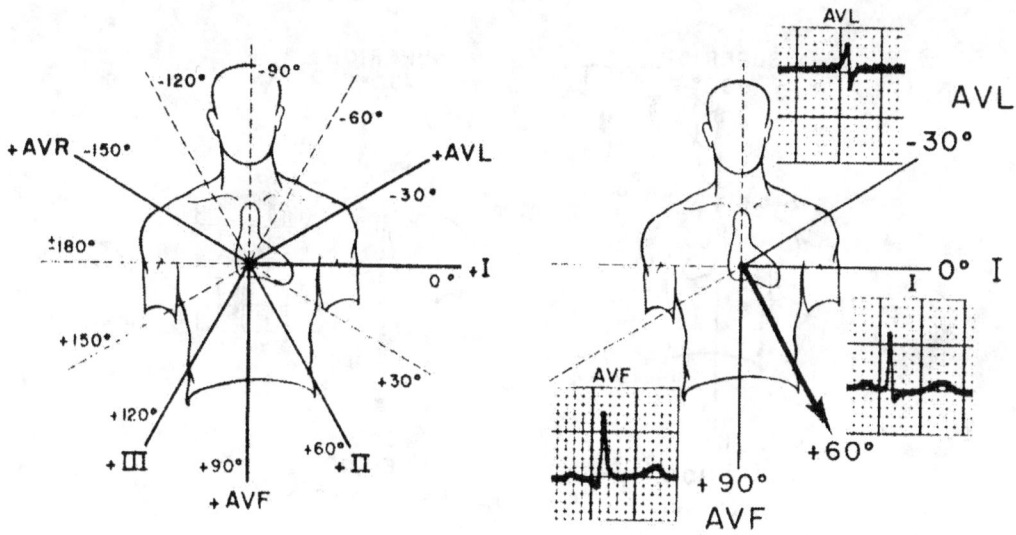

Figure 65

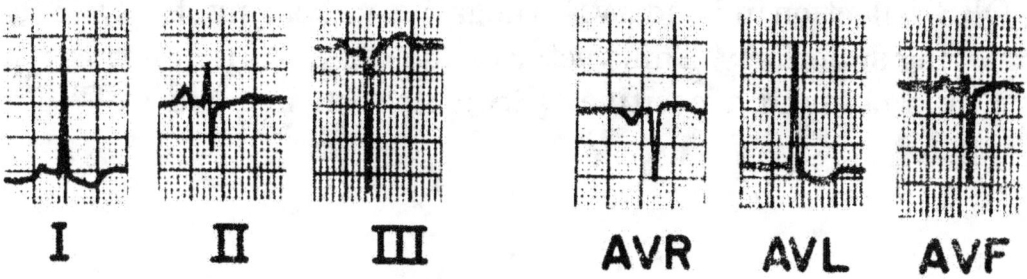

I   II   III   AVR   AVL   AVF

Example 2

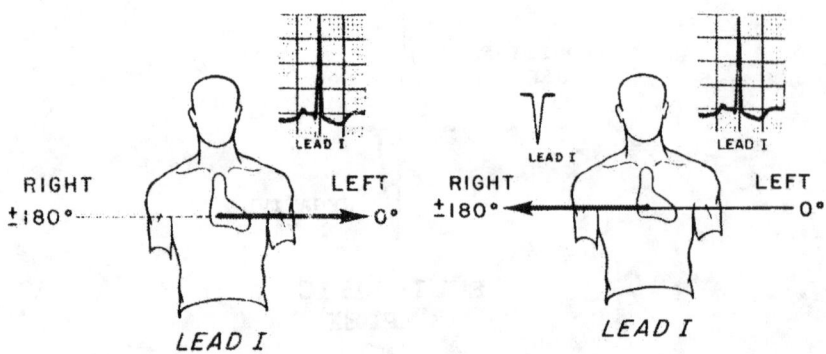

Figure 66

In Example 2 the R wave in Lead I indicates QRS forces are moving to the left. Forces cannot be moving to the right since this would demand a negative QRS in Lead I (Figure 66).

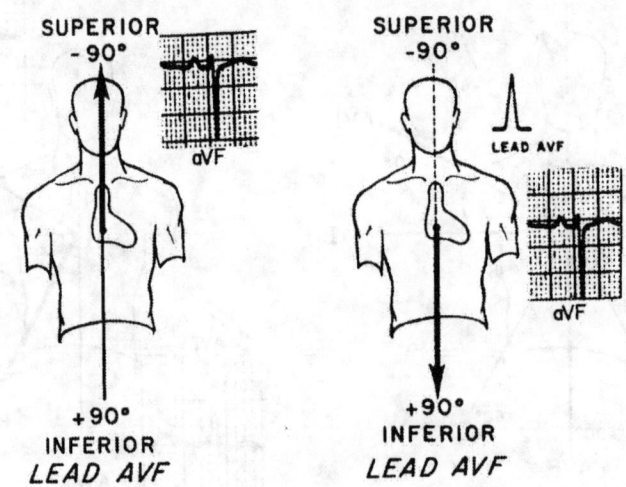

Figure 67

The QRS deflection in Lead aVF is negative indicating that QRS forces are moving superiorly. This vector cannot be moving inferiorly since this would demand a positive QRS deflection in Lead aVF (Figure 67).

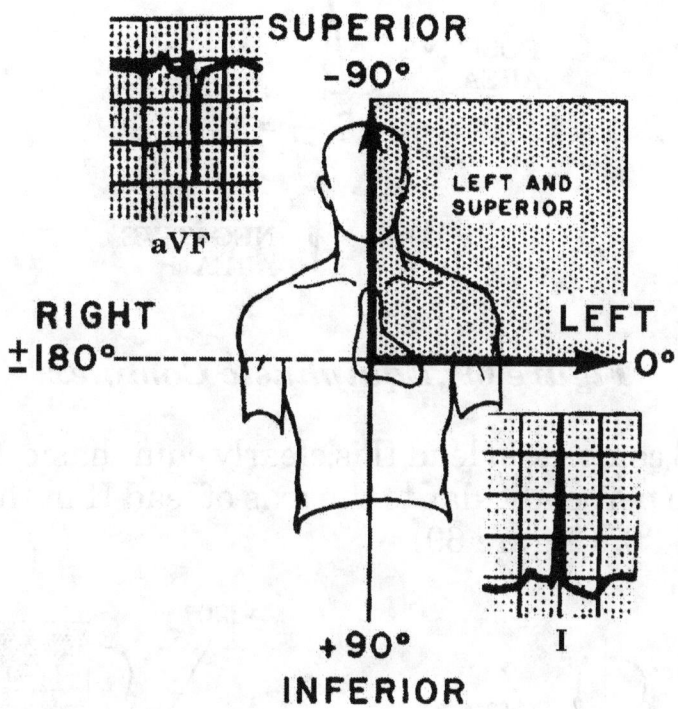

Figure 68

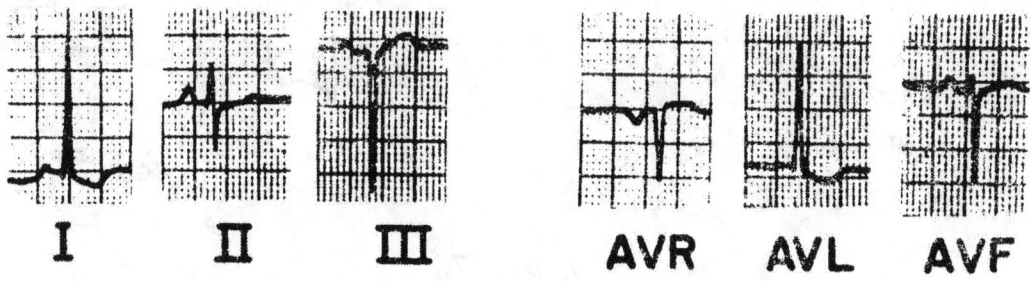

In the above EKG, the R wave in I means QRS forces are moving to the left. The negative deflection in aVF means QRS forces are directed superiorly. The mean QRS vector must lie in the left superior quadrant somewhere between 0° and minus 90° (Figure 68).

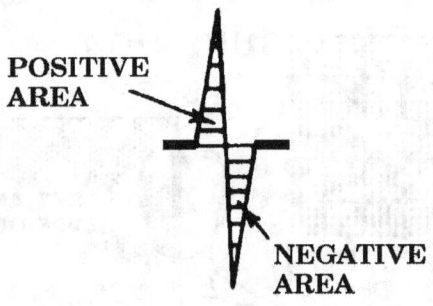

POSITIVE AREA

NEGATIVE AREA

Figure 69: Equiphasic Complex

Since the QRS complex in lead II is clearly equiphasic, the mean QRS vector must be perpendicular to the axis of lead II in the left superior quadrant at —30° (Figure 69).

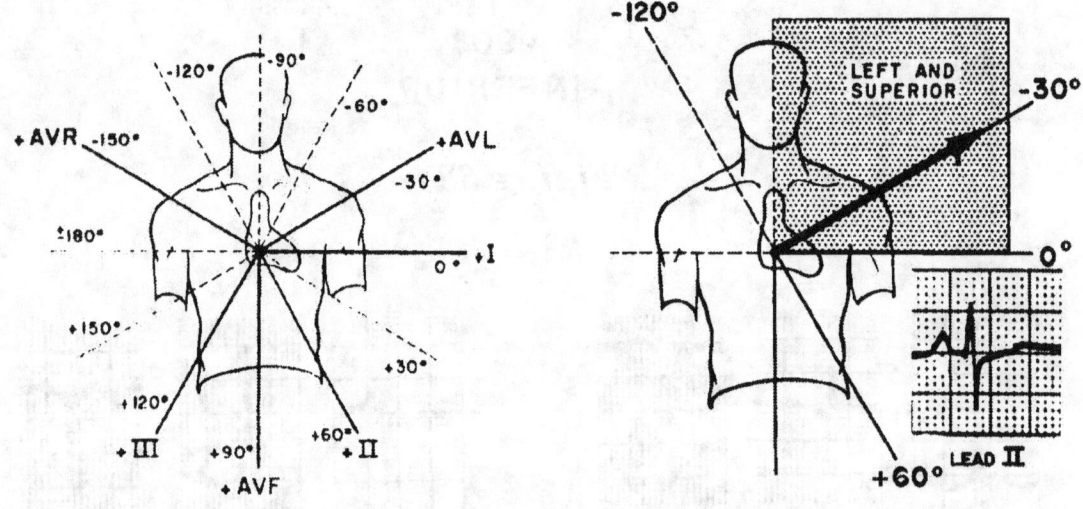

Figure 70

3. Mean Frontal T Wave Vector

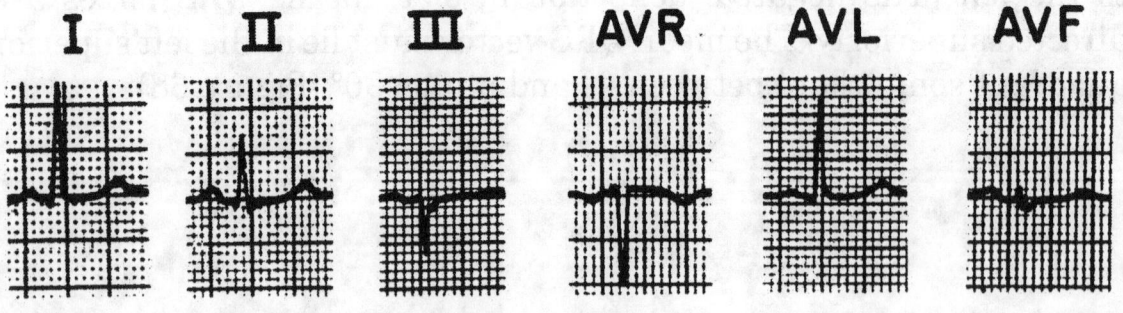

I II III AVR AVL AVF

Example 3

Mean T wave vectors are located in the same manner as mean QRS vectors. In Example 3 the T wave is positive in Lead I indicating that the T vector is directed to the left. The T wave is positive in aVF meaning that the T vector is moving inferiorly. This places the mean T wave vector in the left inferior quadrant somewhere between 0° and +90° (Figure 71).

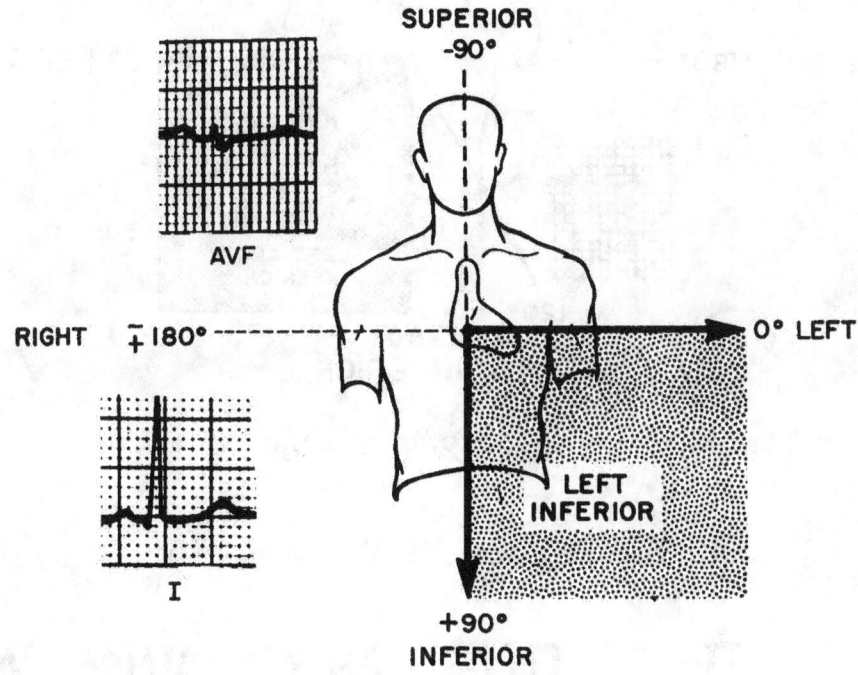

Figure 71

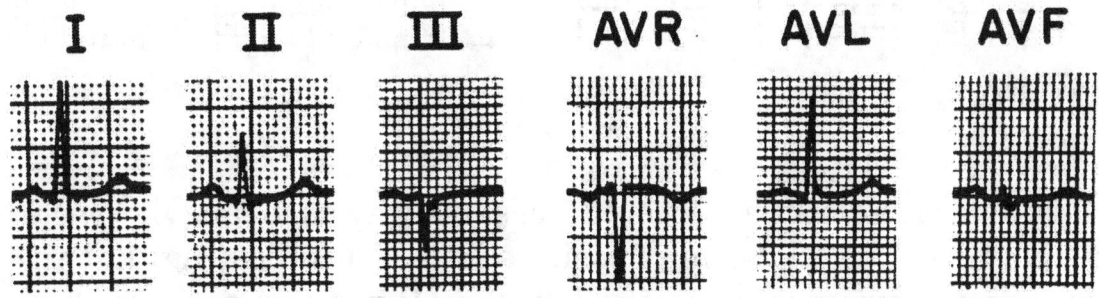

The T wave in lead III is flat and is zero (the same as equiphasic). The T wave vector must lie perpendicular to the axis of lead III in the left inferior quadrant, or +30° (Figure 72).

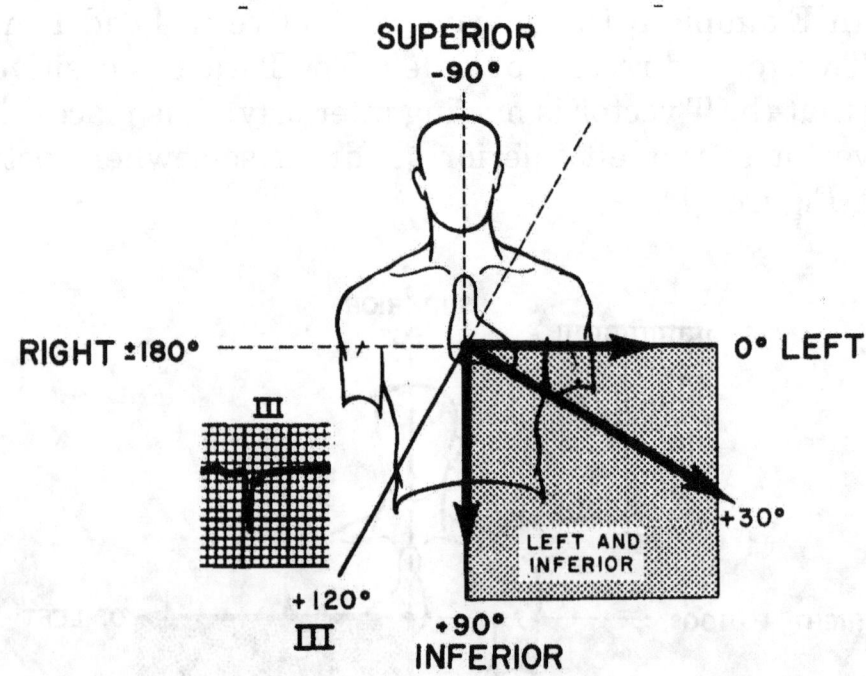

Figure 72: T Vector

4. Mean Frontal P Wave Vector

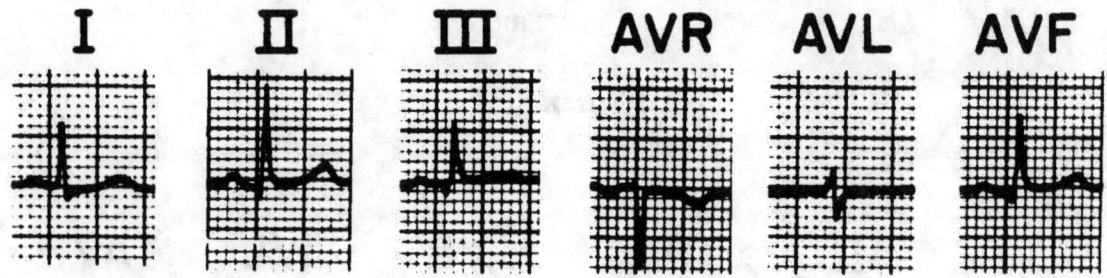

Example 4

Mean P wave vectors are located the same way as mean QRS vectors. In Example 4, the P wave is positive in lead I indicating that the P wave forces are moving to the left. The P wave is positive in aVF meaning that the P wave forces are moving inferiorly. This places the mean P wave vector in the left inferior quadrant somewhere between 0° and +90°. (Figure 73)

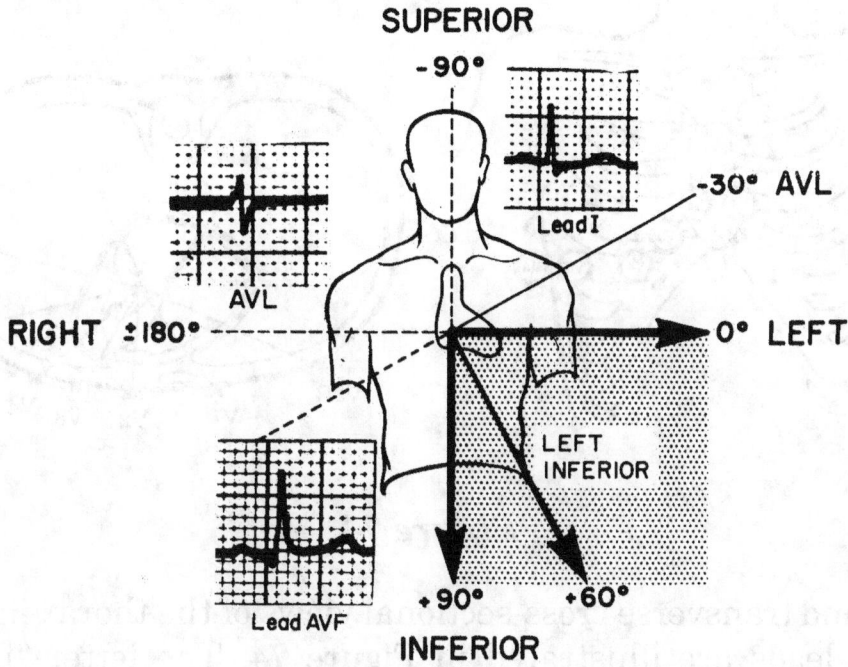

Figure 73: P Vector

The P wave in aVL is flat or zero (the same as equiphasic). The P wave vector must lie perpendicular to the axis of aVL in the left inferior quadrant, or +60° (Figure 73).

D. Mean Transverse Plane Vectors

The positive electrodes of the precordial chest leads V_1 through V_6 are located on the anterior and lateral aspects of the chest. All the precordial leads share the same negative electrode, a central terminal set at zero volts and assumed to be equivalent to the "e" point within the heart. The location of the positive electrodes of the V leads is shown in Figure 74.

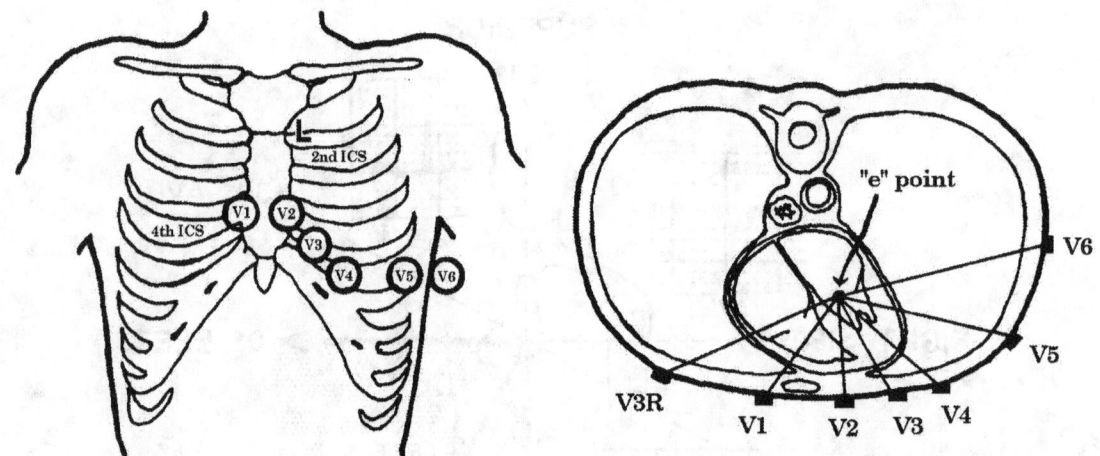

Figure 74

Frontal and transverse cross-sectional views of the thorax in relation to the V leads are illustrated in Figure 74. The letter "L" in the frontal view stands for the "angle of Lewis" found at the junction between the manubrium and body of the sternum. It is a landmark used to locate the second intercostal space, which makes it easier to find the fourth intercostal space. The positive electrodes of leads V_1 and V_2 are in the fourth intercostal space. The positive electrode of lead V_4 is in the fifth intercostal space at the midclavicular line. The positive electrode of lead V_3 is midway between V_2 and V_4. The positive electrode of lead V_5 is at the same level as V_4 at the anterior axillary line. The positive electrode of lead V_6 is at the same level as V_4 and V_5 at the midaxillary line (Figure 75).

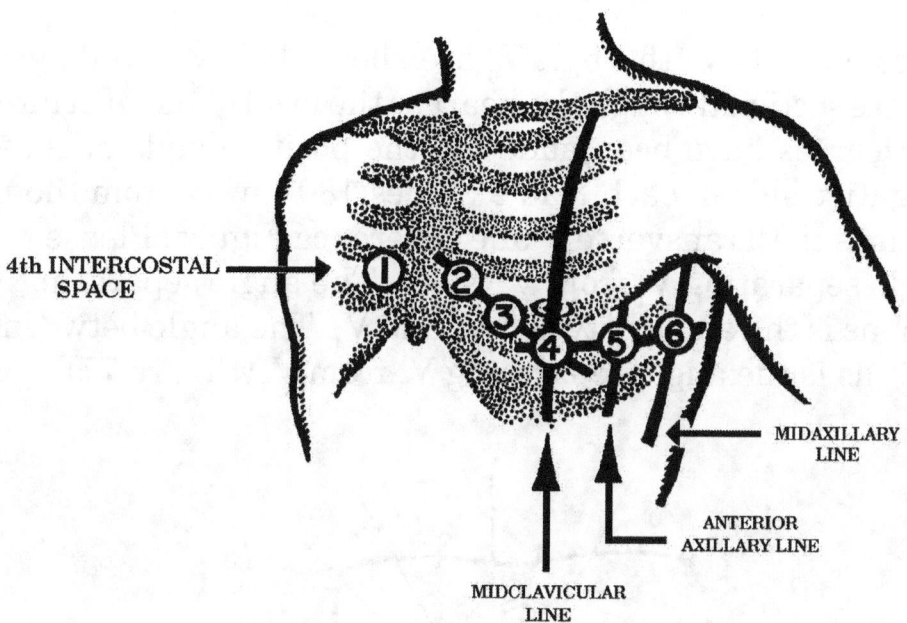

4th INTERCOSTAL SPACE

MIDAXILLARY LINE

ANTERIOR AXILLARY LINE

MIDCLAVICULAR LINE

Figure 75

The precordial leads lie in the transverse plane which divides the body into an upper half and lower half and is defined by the axes of V_2 and V_6 (Figure 76).

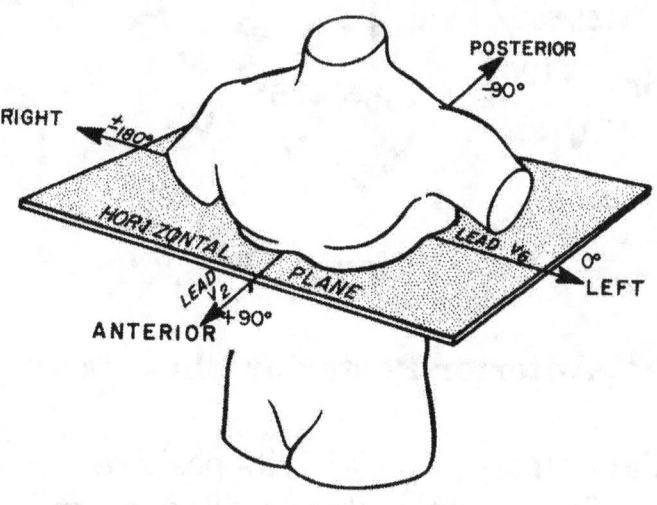

POSTERIOR −90°

RIGHT +180°

HORIZONTAL PLANE

LEAD V_2

LEAD V_6

0° LEFT

ANTERIOR +90°

Figure 76

1. Transverse Plane Reference Figure

The axes of leads V_1 through V_6 are shown below in a diagram of a transverse section through the heart at the level of the fifth intercostal space. Degrees have been added to the positive ends of these axes. The negative side of each lead axis lies 180° away from the positive end. This is the Transverse Plane Reference Figure. Please note that the angle separating V_1 from V_2 is 30°. The angle separating V_2 from V_3 is 15° as is the angle between V_3 and V_4. The angle between V_4 and V_5 is 30° as is the angle separating V_5 from V_6 (Figure 77).

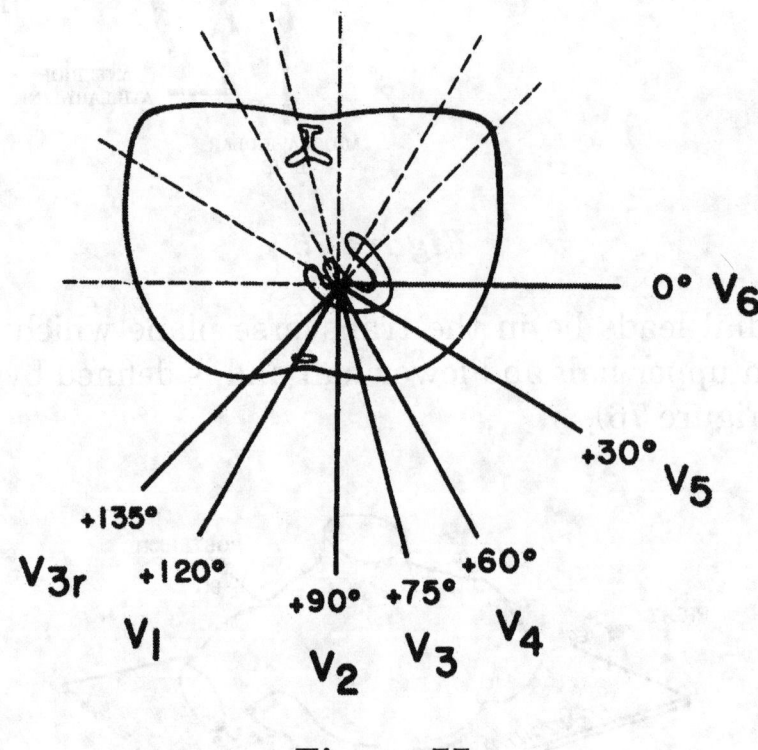

Figure 77

2. Anterior-Posterior Chest Lead

V_2 is a precordial unipolar lead with its positive electrode located at the intersection of the fourth intercostal space and the left sternal edge. Its negative electrode is assumed to be the "e" point of the heart. An imaginary line connecting the positive electrode and the "e" point is the axis of V_2. It has an anterior—posterior orientation (Figure 78).

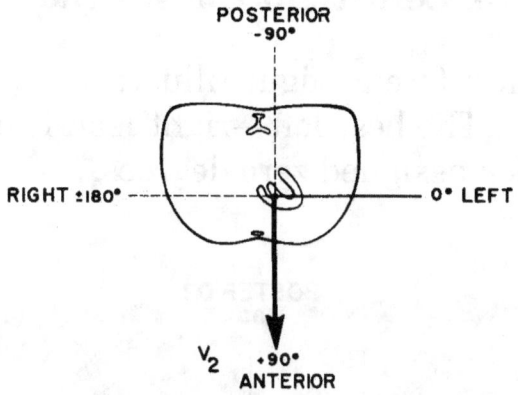

Figure 78

The Quadrant and Perpendicular Rules for the frontal plane have been described. These same rules apply to the transverse plane. The precordial leads may be utilized to determine how far anterior or posterior the frontal vector is shifted. If the deflection in V_2 is positive, the frontal vector must be directed anteriorly (Figure 79 left). If the deflection in V_2 is negative, the frontal vector must be directed posteriorly. (Figure 79 right)

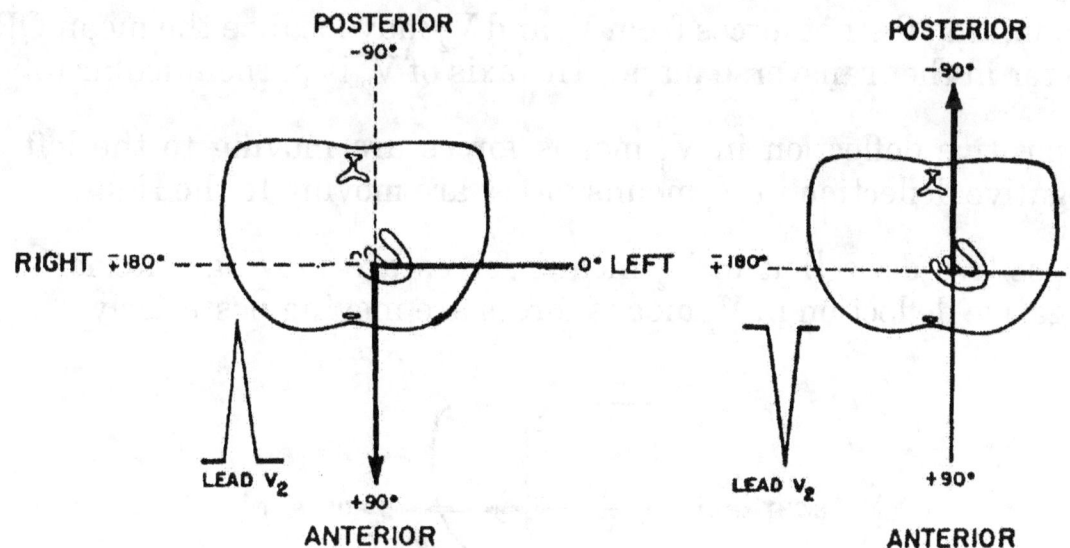

Figure 79

3. Left-Right Chest Lead

The transverse plane reference figure illustrates the axes of precordial leads V_1 through V_6. The best left—right lead is obviously V_6 whose positive end has been assigned zero degrees.

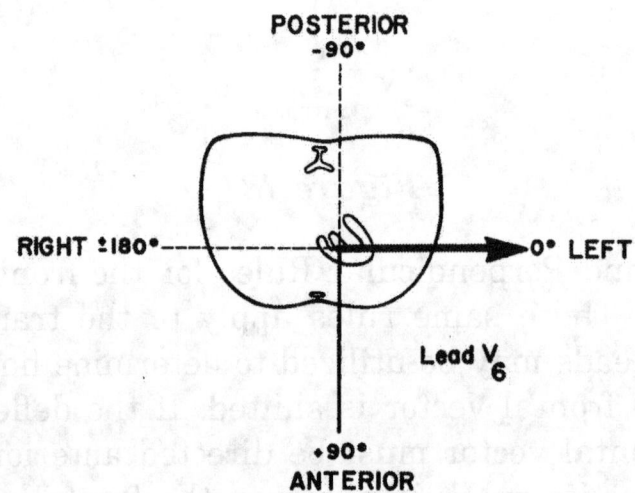

Figure 80

Combining the net forces from V_6 and V_2 may localize the mean QRS vector in the transverse plane. The axis of V_2 is perpendicular to V_6.

A positive deflection in V_6 means forces are moving to the left. A negative deflection in V_6 means forces are moving to the right.

A positive deflection in V_2 means forces are moving anteriorly. A negative deflection in V_2 means forces are moving posteriorly.

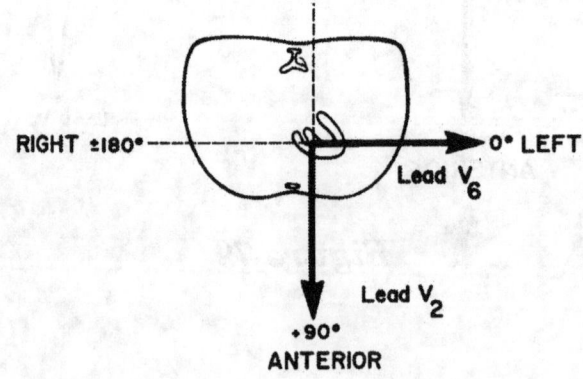

Figure 81

4. Mean Transverse QRS Vector

The mean transverse QRS vector of any electrocardiogram may be located by the Quadrant and Perpendicular Rules of Spatial Analysis. The quadrant rule localizes the mean transverse QRS vector to one of the four transverse plane quadrants. (Figure 82) This is accomplished by looking at the QRS complexes in leads V_6 and V_2. The perpendicular rule states that the mean QRS vector is perpendicular to the lead with an equiphasic or transitional complex and in the preselected quadrant.

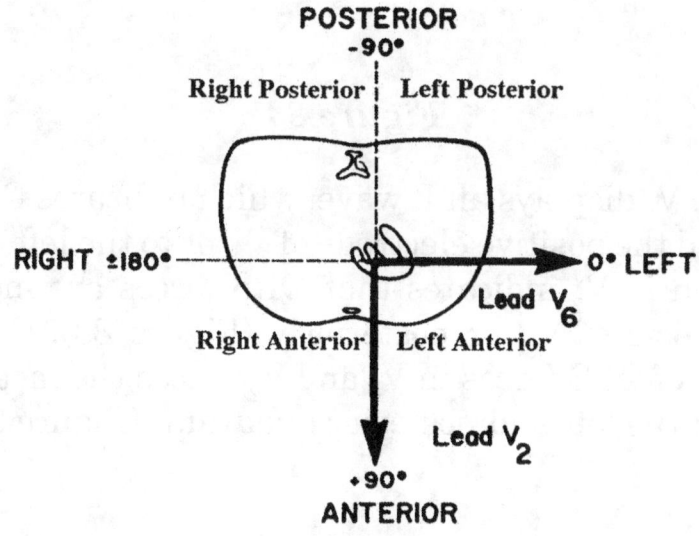

Figure 82

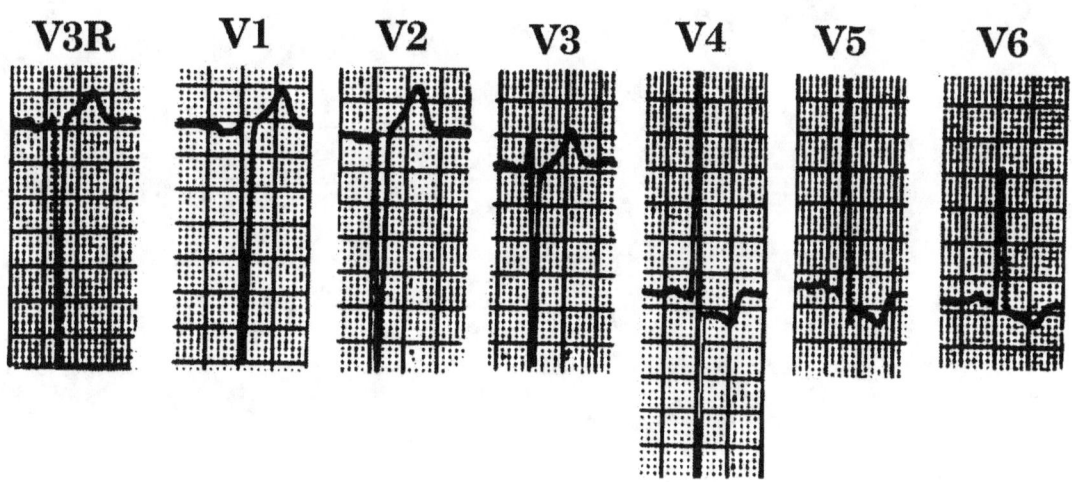

Example 5

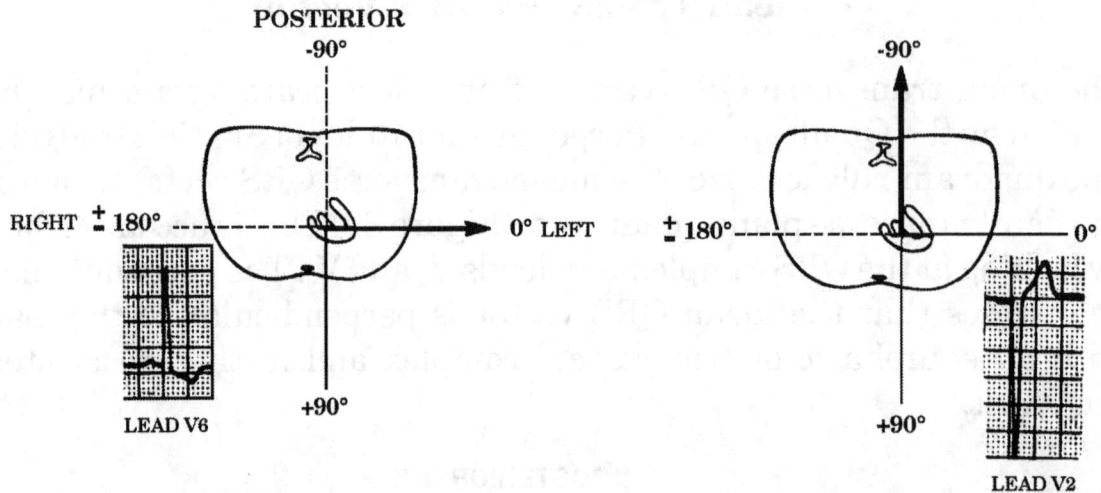

Figure 83

In Example 5, V_6 displays an R wave, which indicates QRS forces are moving toward the positive electrode of V_6, or to the left. The negative QRS deflection in V_2 indicates that QRS forces are moving toward the negative pole of V_2, or posteriorly (Figure 83). Combining the net directions of QRS forces in V_6 and V_2 places the mean transverse QRS vector in the left and posterior quadrant. (Figure 84)

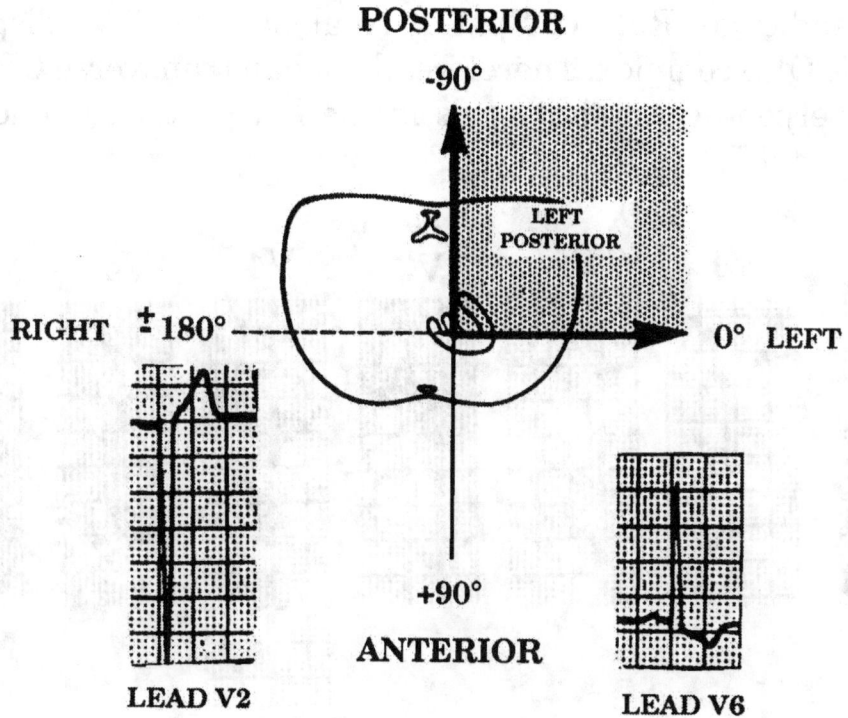

Figure 84

The mean transverse QRS vector cannot be moving to the right since this would demand a negative deflection in V_6. The mean transverse QRS vector cannot be moving anteriorly since this would require a positive deflection in V_2 (Figure 85).

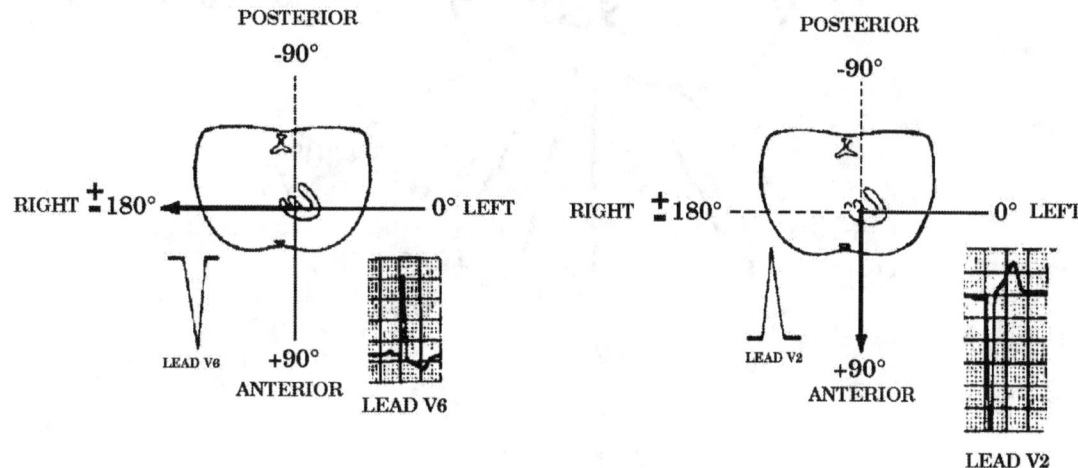

Figure 85

The mean transverse QRS vector may be fixed in degrees using the Perpendicular Rule of Spatial Analysis. Lead V_4 displays an equiphasic QRS complex. Therefore, the mean transverse QRS vector must lie perpendicular to its axis in the left posterior quadrant, at —30° (Figure 87).

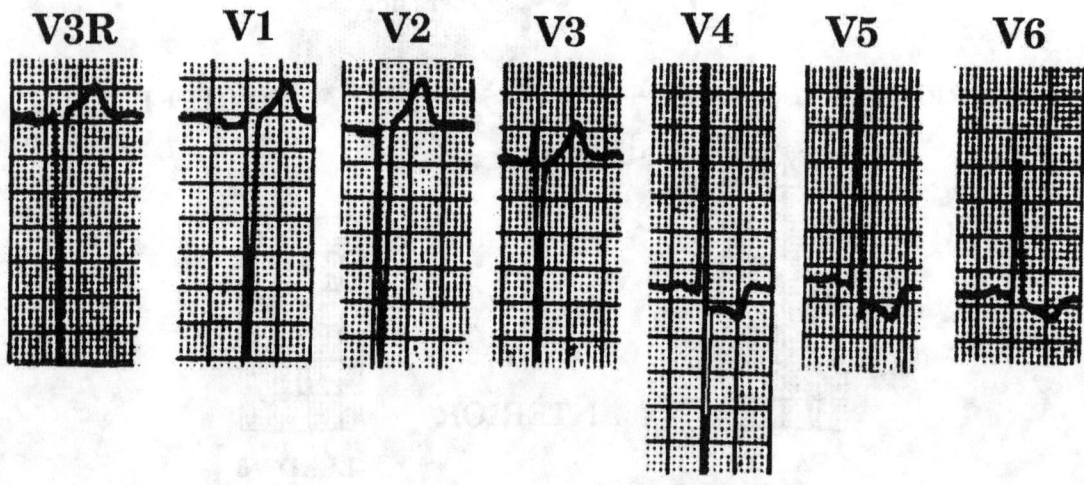

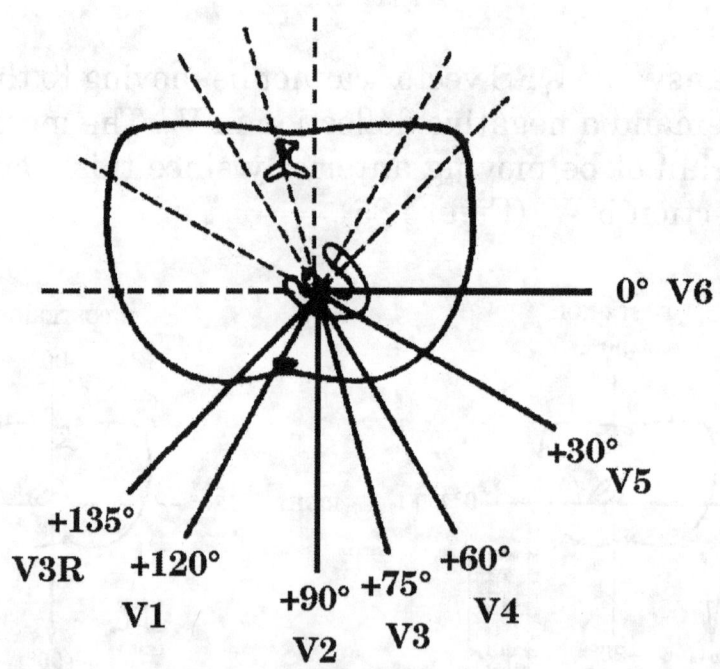

Figure 86

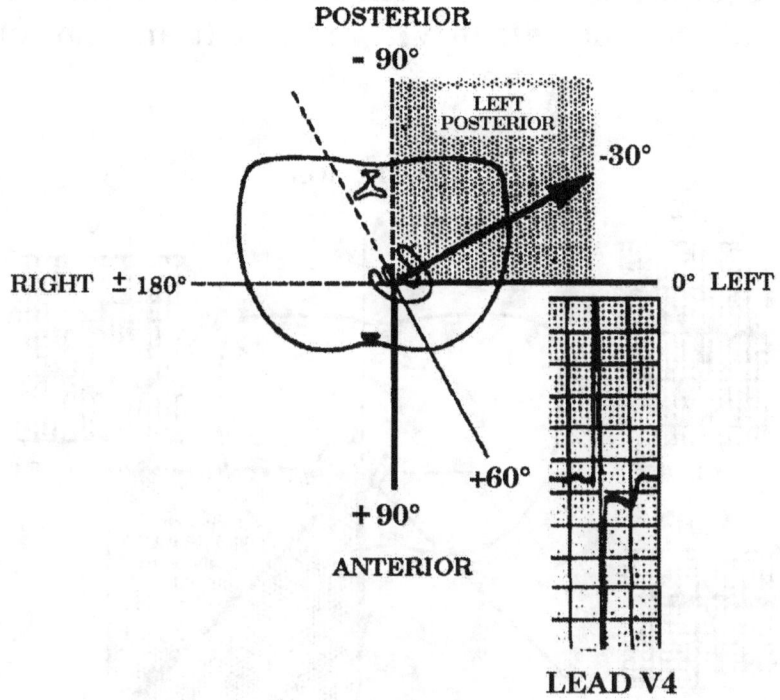

Figure 87

5. Mean Transverse T Vector

The positive T in V_2 indicates that the mean T vector is anterior and the positive T in V_6 means the T vector is directed to the left. The mean transverse T vector, therefore, is left and anterior

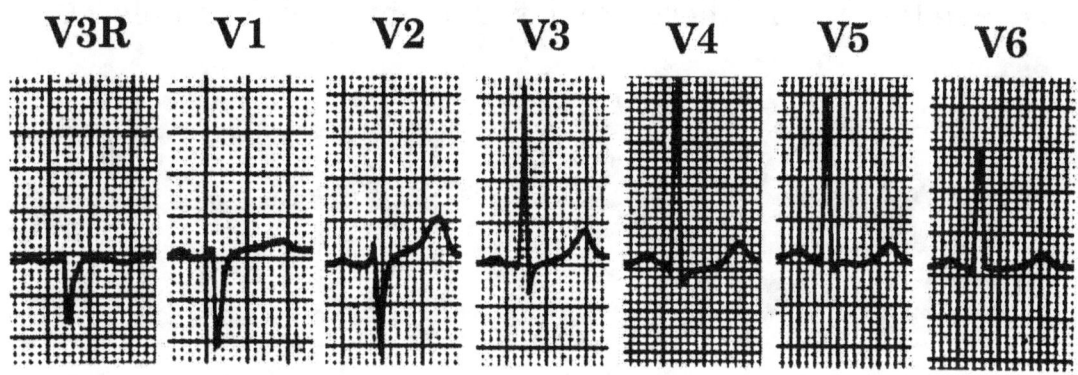

Example 6

T waves are often too small making it necessary to select the lead with the most nearly equiphasic or flat T. In the above tracing, the T most nearly equiphasic is in V$_3$R. The mean transverse T vector must be perpendicular to the axis of V$_3$R in the left anterior quadrant at +45° (Figure 88).

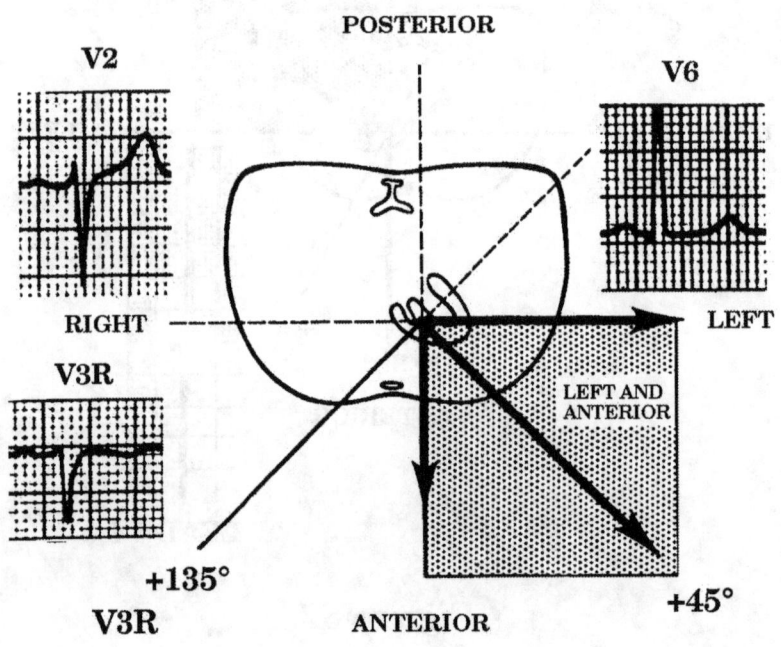

Figure 88

6. Mean Transverse P Vector

The positive P in V$_2$ indicates that the mean P vector is anterior. The positive P in V$_6$ means the P vector is directed to the left. The mean transverse P vector is, therefore, left and anterior (Example 7).

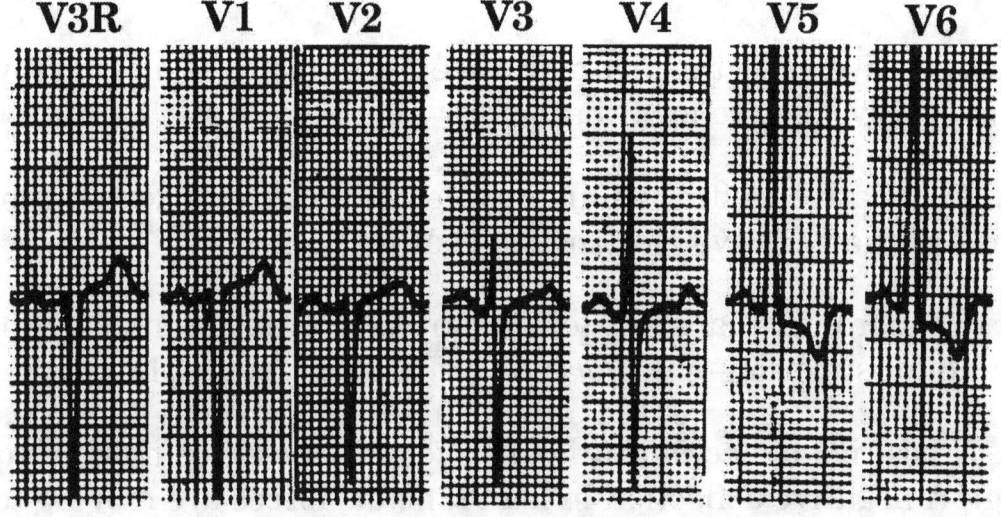

V3R V1 V2 V3 V4 V5 V6

Example 7

The equiphasic P in V_1 fixes the mean transverse P vector left, anterior and perpendicular to V_1 at +30° (Figure 89).

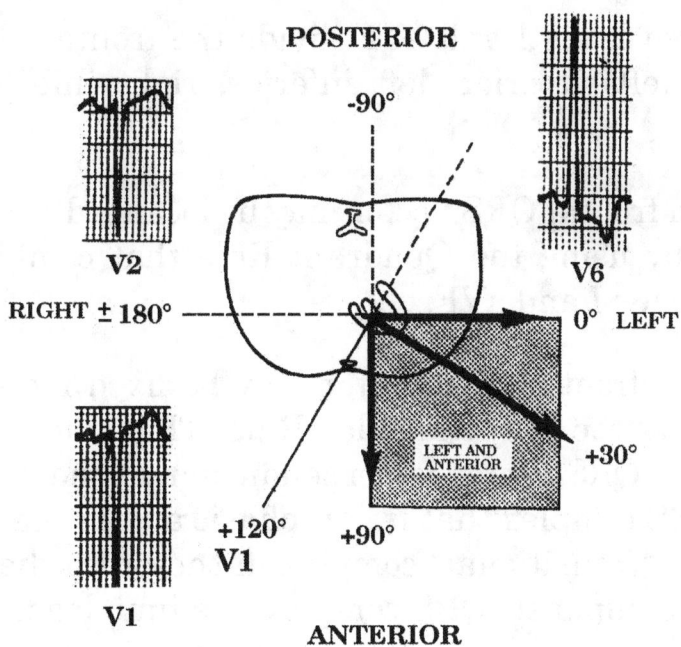

Figure 89: Mean Transverse P Vector

Spatial Analysis Summary

a. A positive deflection in any lead represents forces moving toward the positive electrode of that lead.

b. A negative deflection in any lead represents forces moving toward the negative electrode of that lead.

c. In the frontal plane, for left—right axis orientation look at Lead I and for superior—inferior axis orientation look at lead aVF.

d. A positive deflection in lead I means forces are moving to the left. A negative deflection in lead I means forces are moving to the right.

e. A positive deflection in lead aVF means forces are moving inferiorly. A negative deflection in lead aVF means forces are moving superiorly.

f. The axes of leads I and aVF divide the frontal plane into four quadrants: left superior, left inferior, right inferior, and right superior.

g. The mean frontal QRS vector may be localized to one of the four quadrants by using the Quadrant Rule that combines the QRS vectors in leads I and aVF.

h. The mean frontal QRS vector may be fixed more precisely in degrees using the Perpendicular Rule. This rule states that the mean frontal QRS vector is perpendicular to the axis of the lead with the QRS complex that is equiphasic and in the pre—selected quadrant. A "transitional" complex in the precordial leads is the same as an equiphasic QRS complex in a limb lead.

i. The anterior-posterior axis of the transverse plane is represented best by that of lead V_2.

j. A positive deflection in V_2 represents forces moving anteriorly. A negative deflection in lead V_2 indicates forces are moving posteriorly.

k. The mean frontal QRS vector is shifted anteriorly in the transverse plane when the QRS in V_2 is positive, and posteriorly when the QRS in V_2 is negative.

CHAPTER THREE:
NORMAL EKG

A. Methods and Motives

The technique of localizing the direction of cardiac forces described in Chapter Two will be used throughout the remainder of the manual. Mean vectors of the P wave, QRS complex, and T wave will be determined using the quadrant and perpendicular rules of spatial analysis. Emphasis will be on the mean QRS vector, T vector, the QRS-T angle and electrical axis.

1. Recording Paper

The recording paper used in electrocardiography is divided by a series of horizontal and vertical lines drawn 1 millimeter (mm) apart. They are used to measure deflection amplitude and the duration of the various complexes. The paper speed is exactly 25 mm/sec and as a result each mm along the horizontal time axis represents 1/25 or 0.04 sec. (Figures 1 & 2).

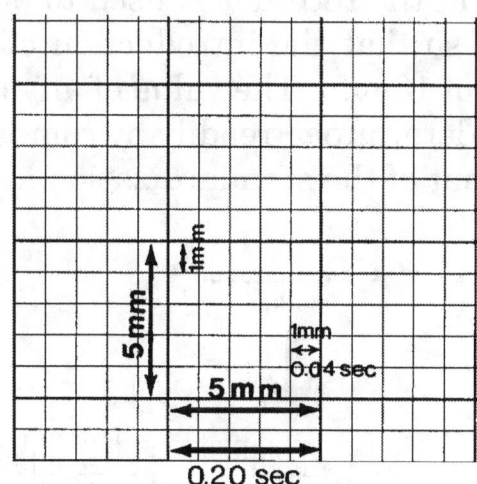

Figure 1

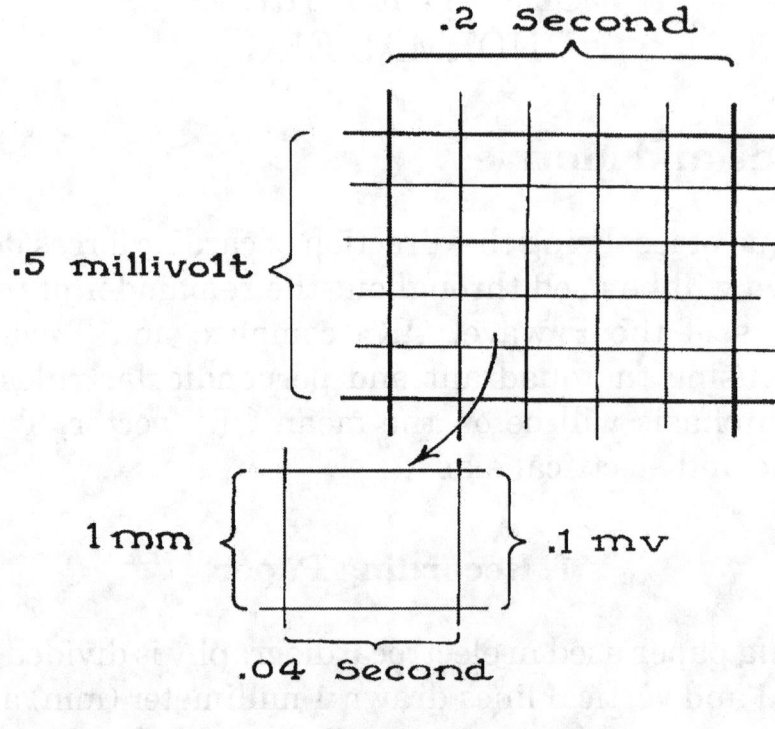

Figure 2

2. Standardization of the Electrocardiogram

A calibration device in the recorder is used to adjust the sensitivity of the galvanometer so that the introduction of a 1 mV signal will produce a deflection of 10 mm. The value of any deflection expressed in millivolts may be determined readily by comparing the amplitude of the deflection to that of the standardization artifact (Figure 3).

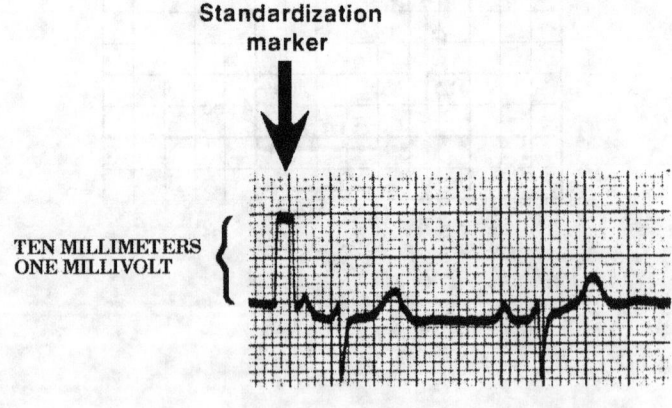

Figure 3

3. Directions and Deflections

The *Einthoven Rule of Leads* (term coined by author) states that a positive (upright) deflection in any lead represents forces moving toward the positive electrode of that lead. A negative (downward) deflection in any lead represents forces moving toward the negative electrode of that lead. (Figure 4).

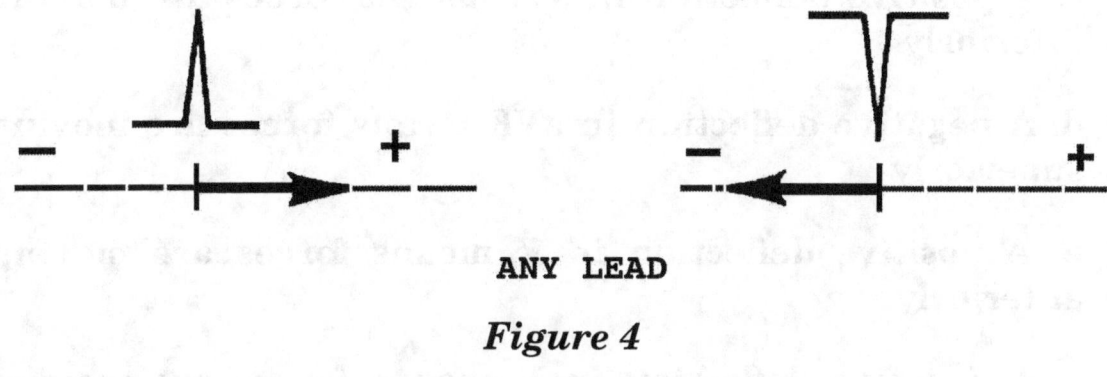

ANY LEAD

Figure 4

THEREFORE,

A **positive** deflection in **lead I** means forces are moving toward its positive electrode on the left arm and away from its negative electrode, or **to the left**. A **negative** deflection in **lead I** means forces are moving toward its negative electrode on the right arm and away from its positive electrode, or **to the right**.

A **positive** deflection in **aVF** means forces are moving toward its positive electrode at the symphysis and away from its negative electrode at the "e" point of the heart, or **inferiorly**. A **negative** deflection in **aVF** means forces are moving toward its negative electrode at the "e" point of the heart and away from its positive electrode, or **superiorly.**

A **positive** deflection in V_2 means forces are moving toward its positive electrode at the fourth intercostal space just to the left of the sternal edge and away from its negative electrode the "e" point of the heart, or **anteriorly**. A **negative** deflection in V_2 means forces are moving toward its negative electrode at the "e" point of the heart and away from its positive electrode, or **posteriorly**.

4. Summary

a. A positive deflection in I means forces are moving to the left

b. A negative deflection in I means forces are moving to the right

c. A positive deflection in aVF means forces are moving inferiorly

d. A negative deflection in aVF means forces are moving superiorly

e. A positive deflection in V_2 means forces are moving anteriorly

f. A negative deflection in V_2 means forces are moving posteriorly

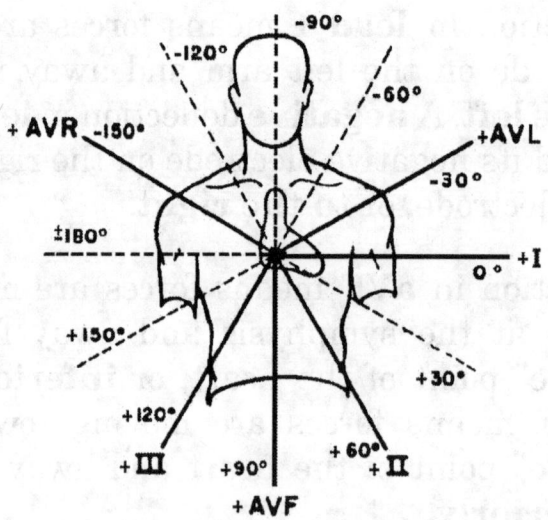

Figure 5

The Hexaxial Reference Figure illustrated above demonstrates the bipolar and unipolar limb leads and their respective axes in degrees. It plays an important role in Spatial Analysis and should be used with the Quadrant and Perpendicular Rules.

5. Quadrant and Perpendicular Rules

With the quadrant rule of spatial analysis, it can be determined in which of the four frontal plane quadrants the mean QRS or T vector is located. With the perpendicular rule of spatial analysis, it is possible to determine the angle in degrees at which the mean QRS or T vector is located. The perpendicular rule states that the mean QRS or T vector, in the preselected quadrant, lies perpendicular to the axis of the lead with the equiphasic complex (Figure 6).

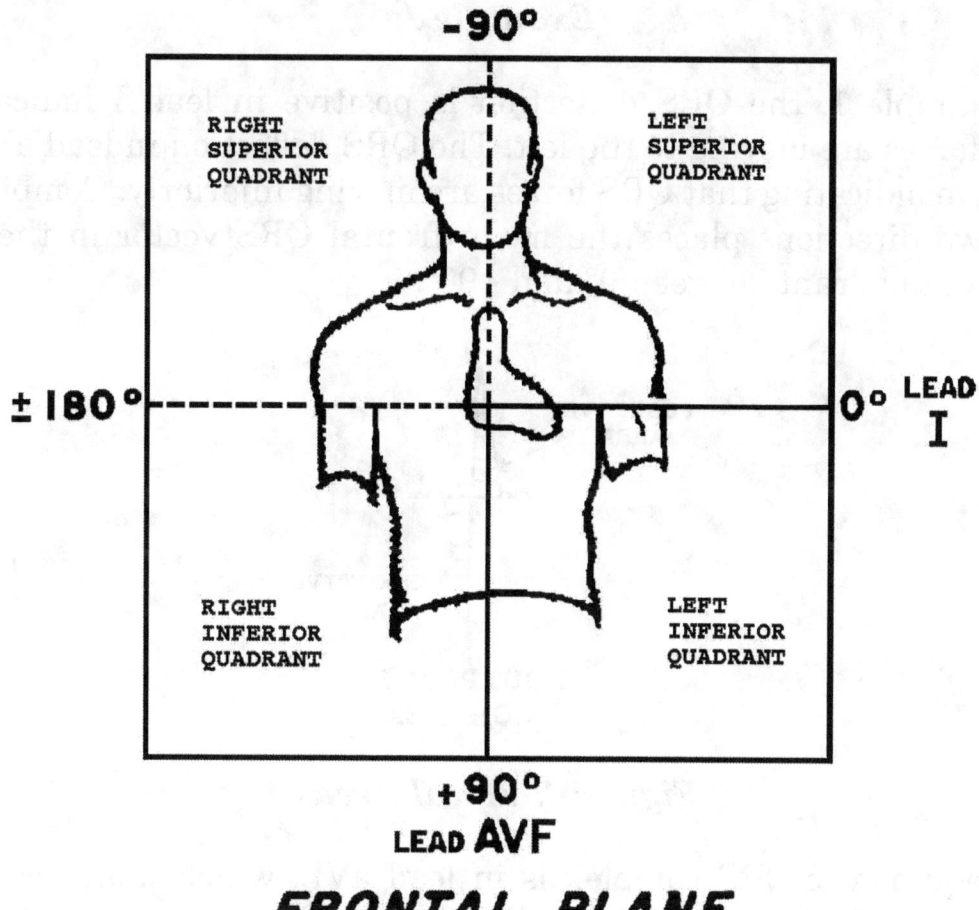

FRONTAL PLANE

Figure 6

6. Mean QRS Vector

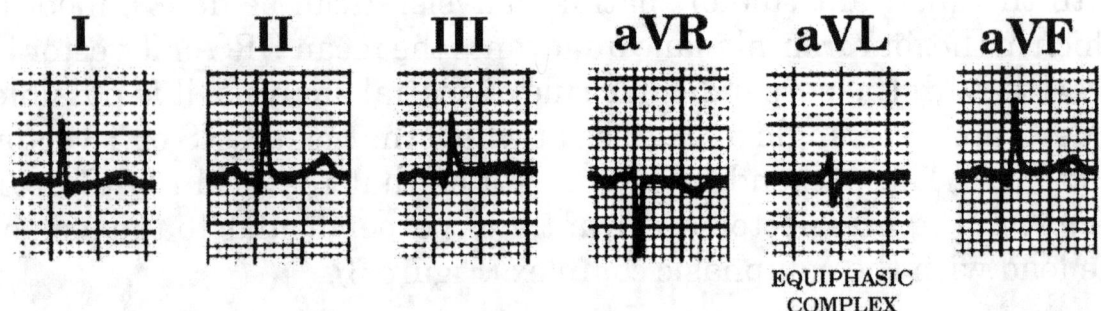

Example 1

In Example 1, the QRS deflection is positive in lead I indicating QRS forces are moving to the left. The QRS deflection in lead aVF is positive indicating that QRS forces are moving inferiorly. Combining the two directions places the mean frontal QRS vector in the left inferior quadrant between 0° and +90°.

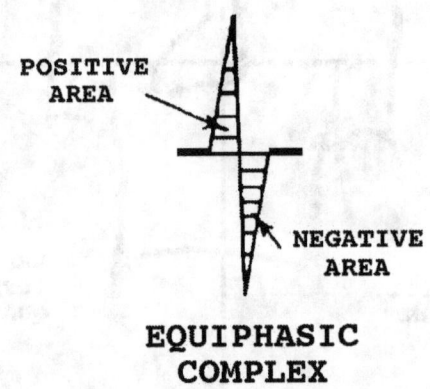

Figure 7: Equal Areas

The equiphasic QRS complex is in lead aVL, which indicates that the mean QRS vector is perpendicular to the axis of aVL in the left inferior quadrant.

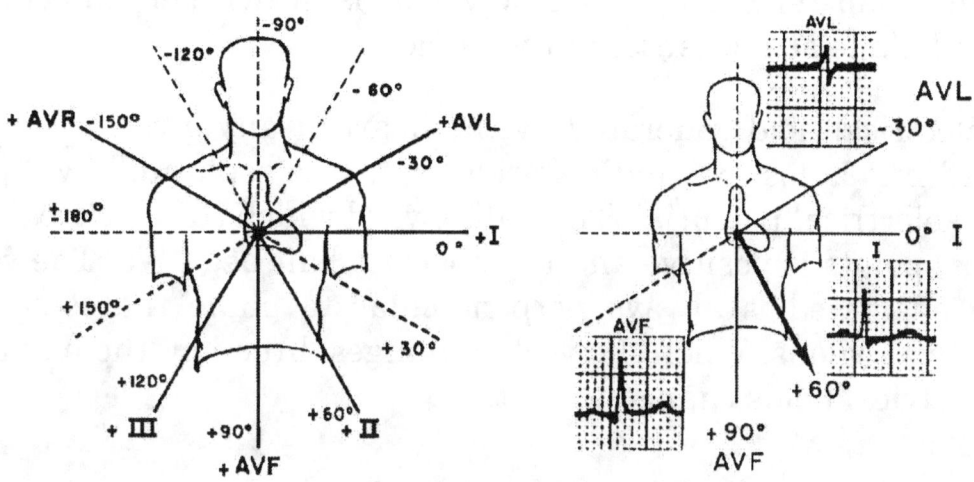

Figure 8

The axis of lead II is perpendicular to the axis of aVL. The angle assigned to lead II in the left inferior quadrant is +60°. Therefore, the mean frontal QRS vector is located at +60° (Figure 8).

7. Null or Transitional Plane

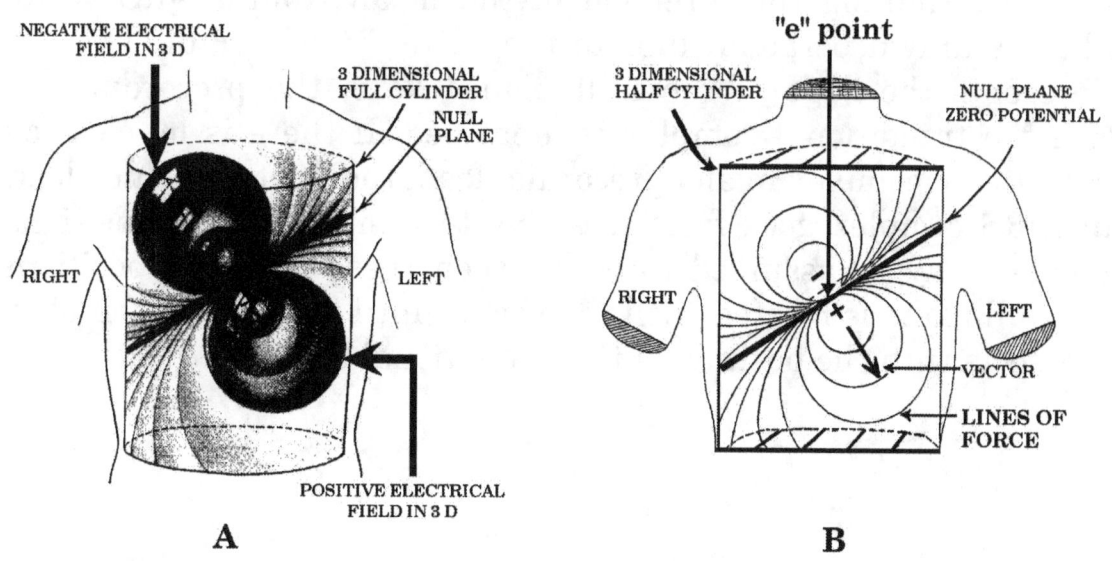

Figure 9

Figure 9A illustrates a three dimensional view of the electrical fields produced by the heart and the cylinder that contains them. Figure 9B shows the cylinder, the electrical fields and the null plane with the front half removed. A cardiac vector is drawn perpendicular to the null plane at the center of the plane.

The electrical field generated by the heart at any given moment is divided into a positive and negative half. It is separated by a plane of zero electrical potential. The half toward which the electrical force (vector) points is positive and the opposite half negative. The vector being evaluated is always perpendicular to plane that separates these two regions. When the vector changes direction, the plane and the electrical fields change with it.

8. Mean Spatial QRS Vector

To determine how far the mean frontal QRS vector is shifted away from the frontal plane, the "transitional" or equiphasic QRS complex must be identified in one of the precordial leads. It is not necessary to localize the anterior or posterior shift of the vector in degrees. Anterior and posterior shifts may be described as "slight", "moderate", or "marked".

After determining the direction of the mean frontal QRS vector, a line is drawn perpendicular to the vector at its origin. This line represents the edge of the null plane. Next, the precordial lead with the transitional complex is identified. If there is no clear-cut transitional complex in any precordial lead, then note in which leads the QRS complex goes from negative to positive. The transitional complex may be assumed to be between these two lead positions. The null plane is rotated with the vector until the edge of the plane intersects with the position of the transitional complex.

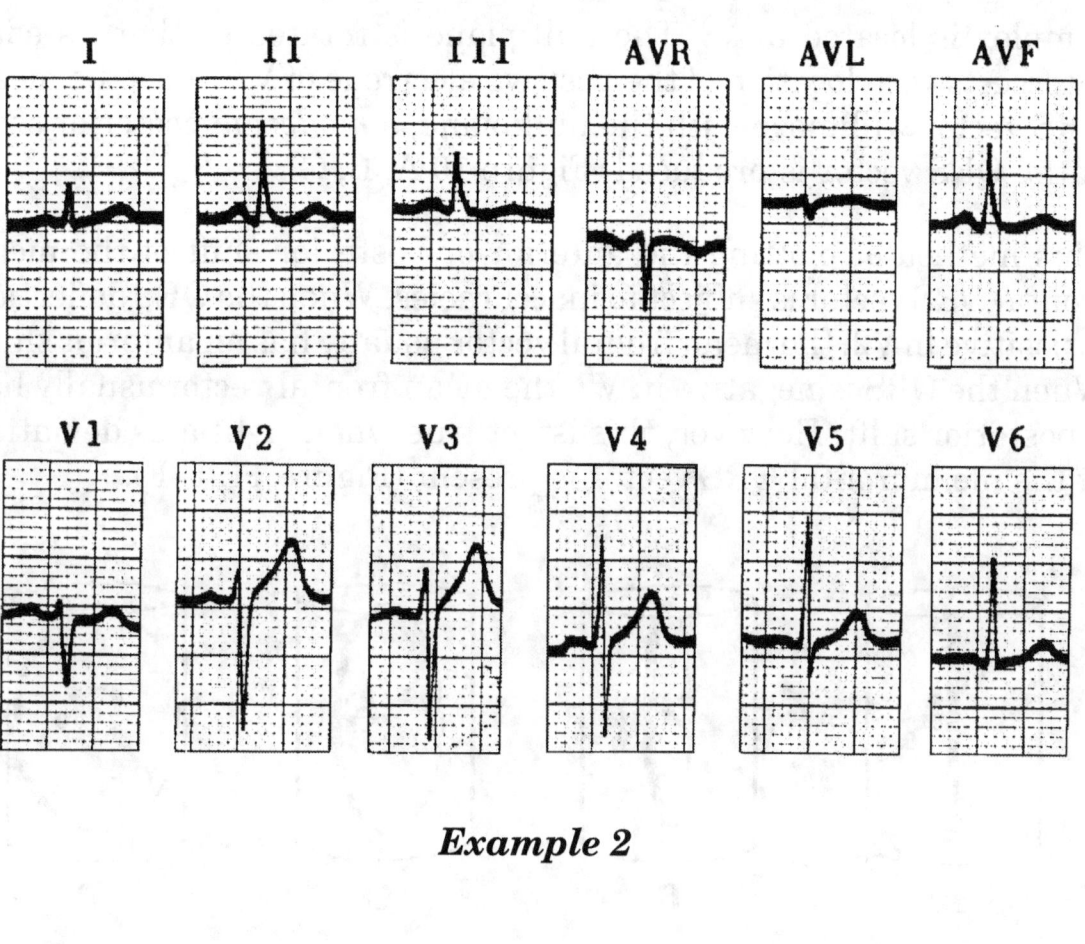

Example 2

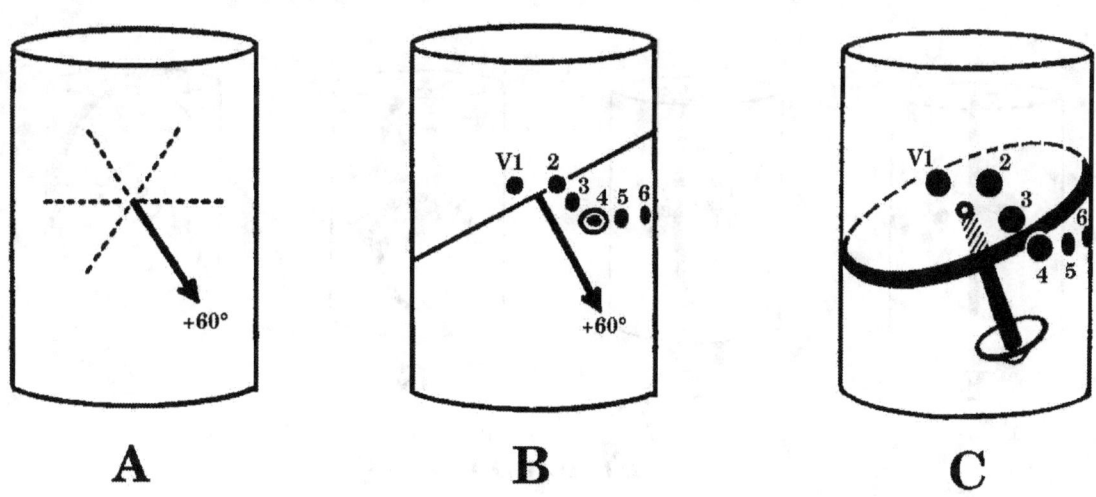

A   **B**   **C**

Figure 10

In Example 2 there are positive deflections, R waves, in leads I and aVF. This places the mean QRS vector in the left inferior quadrant. The QRS in aVL is equiphasic indicating the mean QRS vector is perpendicular to the –30° axis of aVL, or +60°. The transitional QRS complex is located in V_4. The null plane is rotated so that its edge intersects with location of the positive electrode of V_4. Since the mean QRS vector will move with the null plane, a *moderate* posterior shift of the QRS vector is produced (Figure 10 A, B, C).

This method of plotting the anterior or posterior shift of the mean frontal QRS vector is subject to inaccuracies. When the QRS deflection is positive in V2, the mean frontal vector usually has an anterior shift. When the QRS is negative in V2, the mean frontal vector usually has a posterior shift. However, this is not true when right axis deviation of the mean frontal QRS vector is present (Figure 11, E-F).

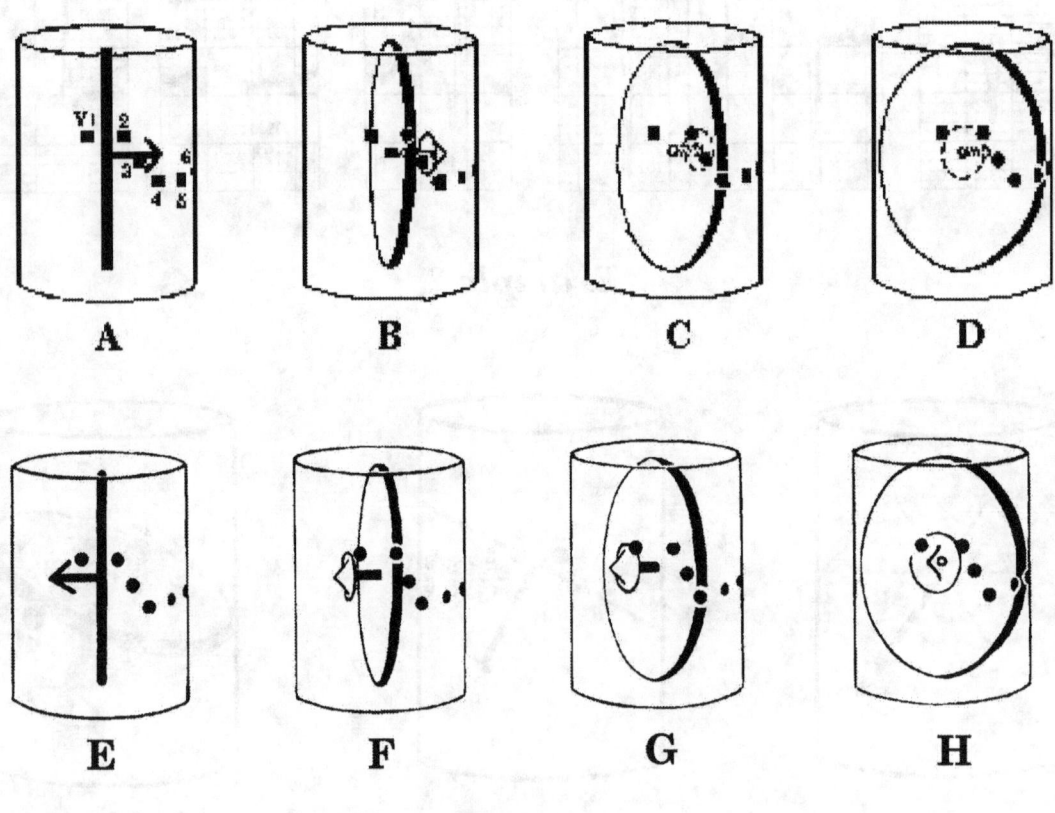

Figure 11

In Figure 11, the thorax is represented by a series of cylinders. The arrows indicate the direction of the QRS vector, and the black dots represent the location of the positive electrodes of the precordial leads. In Figure 11A, the V leads are labeled. The circle within the cylinder is the "null plane". Where its bold edge intersects with a dot, or between two dots, indicates the location of the transitional or equiphasic QRS complex in the precordial V leads.

The cylinders in Figure 11 illustrate two vectors, one shown in A through D, which is directed, to the left at 0°. The other in E through H is directed to the right at 180°. The vectors are drawn perpendicular to the center of the "null" plane. In Figure 11A, the vector is directed leftward and parallel with the frontal plane. This means that there is neither an anterior nor a posterior tilt. The transitional complex is located between V_1 and V_2. It will be recalled that the transitional complex is synonymous with equiphasic complex. In Figure 11B, the transitional QRS complex is shown between V_2 and V_3. In this situation the QRS vector is shifted posteriorly. Figure 11C indicates the transitional complex in lead V_4, and the vector is directed a bit more posteriorly. 11D shows the transitional complex between V_5 and V_6 and the vector is shifted even more posteriorly.

In 11E, the vector is directed rightward and is parallel to the frontal plane. The vector is deviated neither anteriorly nor posteriorly. The transitional complex is located between V_1 and V_2. Figure 11F shows the transitional complex between V_2 and V_3, but the vector is deviated anteriorly and not posteriorly as illustrated in Figure 11B. Figure 11G indicates that the transitional complex is at V_4 and the vector is shown more anteriorly oriented. 11H shows the transitional complex between V_5 and V_6 and the vector is shifted even more anteriorly.

The direction the QRS vector is determined primarily by the dominant ventricular muscle mass. Electrical integrity and anatomical location of the ventricles also play a role. In the normal adult, the predominant muscle mass is the left ventricle. The location of the left ventricle in the thorax is left and posterior. The mean QRS vector will be directed leftward and posteriorly. **Normally, the mean T wave vector lies within 45° to 60° of the mean QRS vector.**

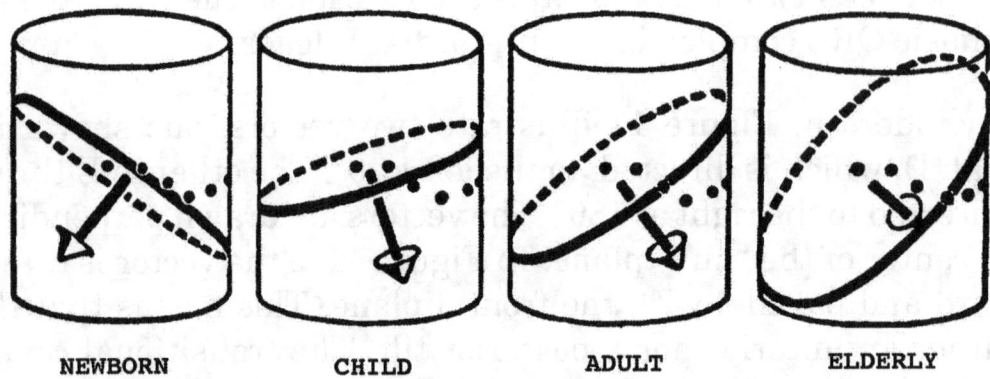

NEWBORN   CHILD   ADULT   ELDERLY

Figure 12

In the neonate, the right ventricle is dominant and is located in the thorax anteriorly and to the right. The mean QRS vector is oriented rightward, anterior and inferior. Over time, the left ventricle becomes more and more dominant causing a leftward, superior and posterior shift of the mean QRS vector (Figure 12).

9. Normal Electrocardiogram

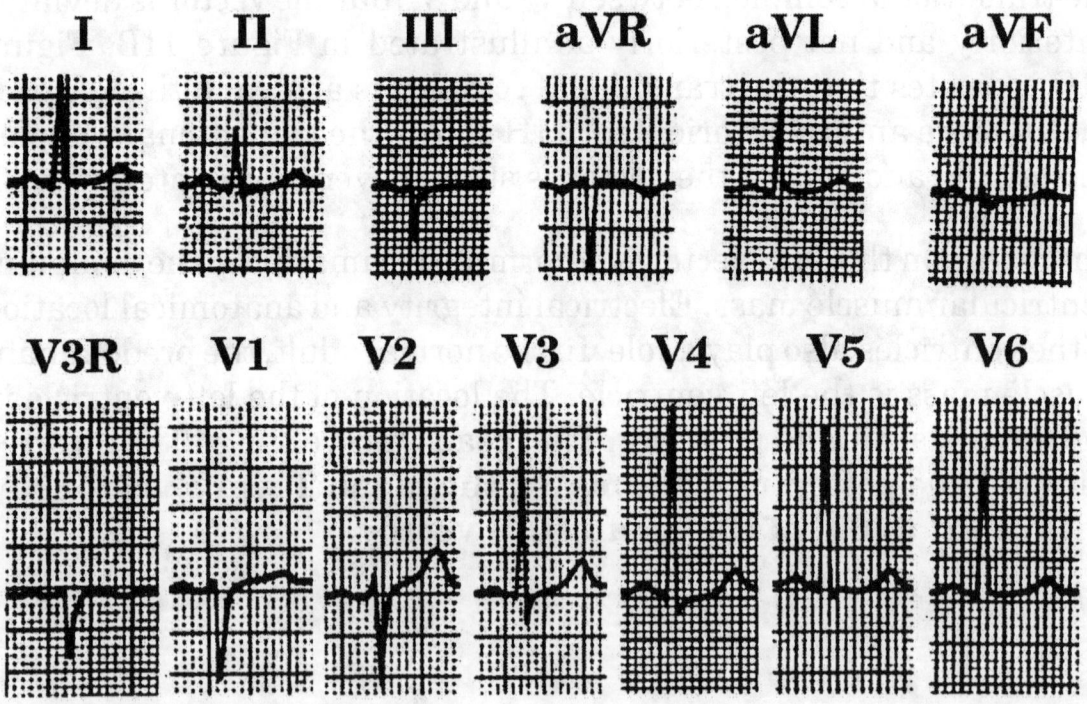

Example 3: Normal EKG

In the electrocardiogram illustrated in Example 3, there is an R wave in I indicating forces are directed to the left. The QRS complex is equiphasic in aVF indicating that that the QRS vector is perpendicular to the axis of lead aVF at 0°.

The QRS in V_2 is negative which means the mean frontal QRS vector, located at 0°, is shifted posteriorly. The QRS complex is negative in V_2 and becomes positive in V_3. The transitional QRS complex must lie between V_2 and V_3. To see how far posterior the mean frontal QRS vector is shifted, rotate the null plane so that its anterior edge intersects with a point midway between V_2 and V_3 (Figure 13).

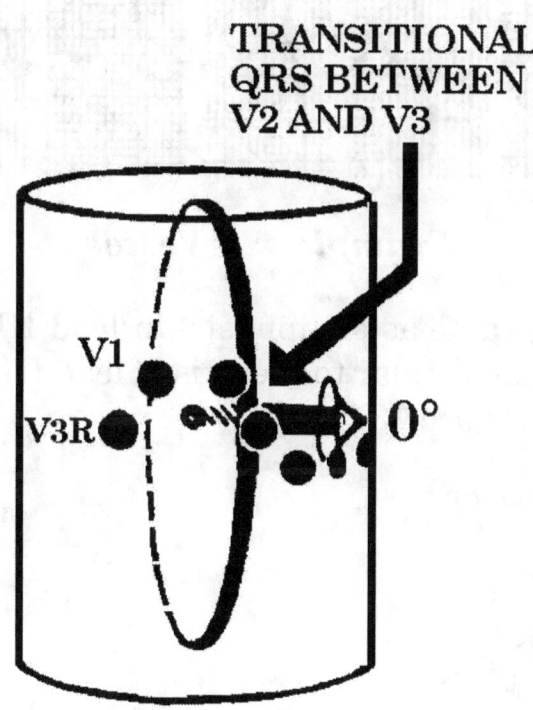

TRANSITIONAL QRS BETWEEN V2 AND V3

Figure 13: Mean Spatial QRS Vector

10. Mean Frontal T Vector

In Example 4, the T is positive in lead I indicating that the T vector is oriented to the left. The T is positive in aVF indicating that the T vector is directed inferiorly. The mean T vector has to lie in the left inferior quadrant between 0° and +90°.

91

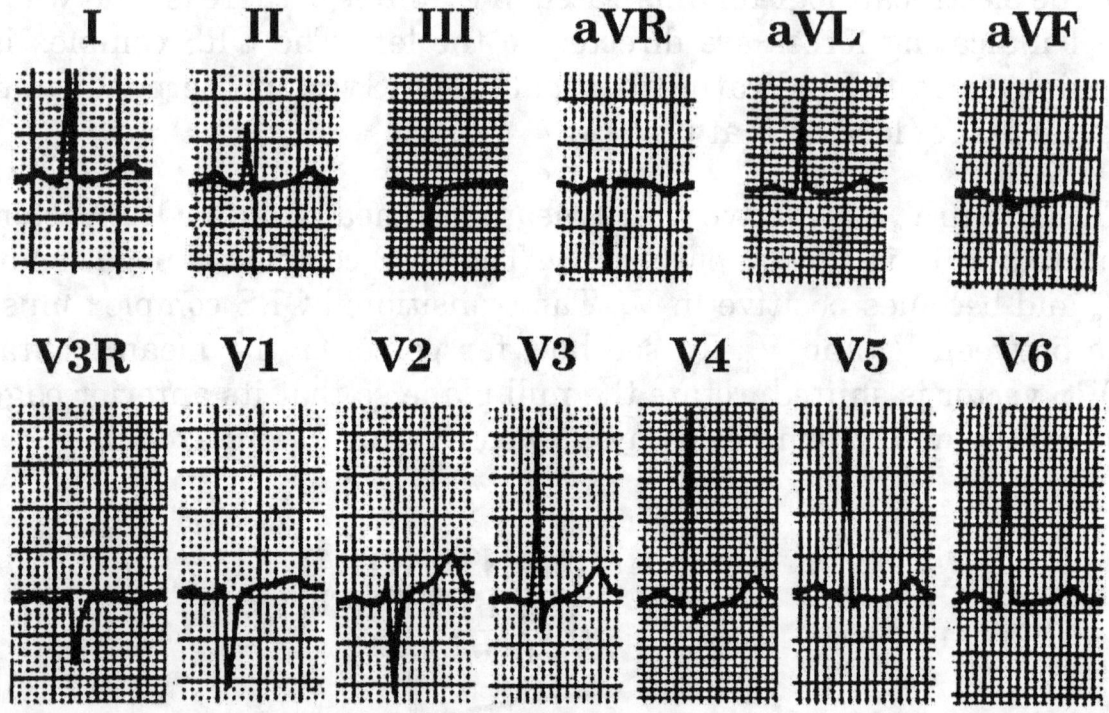

Example 4: T Vector

The T is flat (zero, same as equiphasic) in lead III. Therefore, the T vector must be perpendicular to the axis of lead III in the left inferior quadrant at +30° (Figure 13).

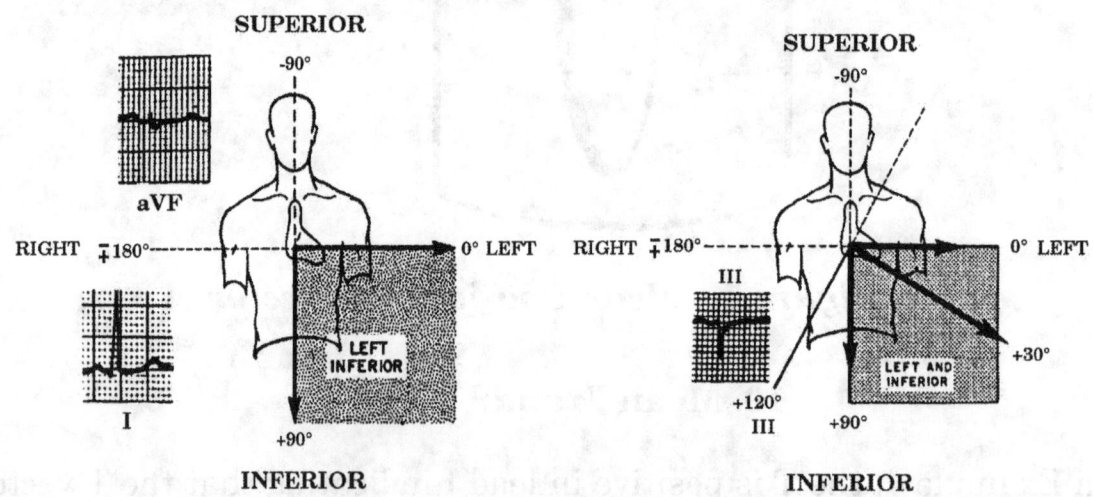

Figure 13: Mean T vector

In Example 4, the T in V_2 is positive indicating that the mean frontal T vector is shifted anteriorly. The T in V_3R is negative but positive in

V_1. The transitional (equiphasic) T wave, therefore, must lie between V_3R and V_1. To determine how far the frontal T vector is shifted anteriorly, move the edge of the null plane so that it intersects with a point midway between V_3R and V_1 (Figure 14).

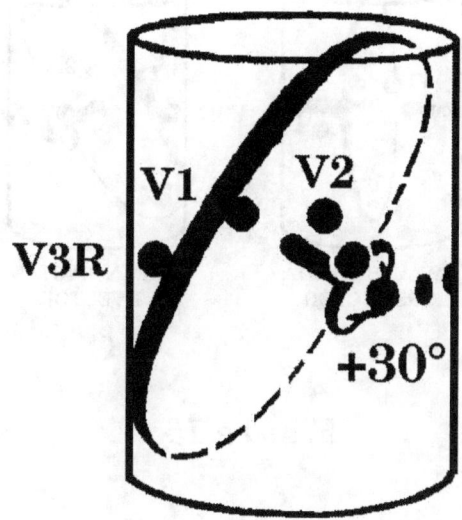

Figure 14: Mean Spatial T Vector

11. QRS-T Angle

The angle between the mean QRS and T vectors represents the relationship between ventricular depolarization and repolarization. It is considered a reliable way to determine if a T wave is normal or abnormal. ***This angle doesn't often exceed 45° in the frontal plane or 60° in the transverse plane.*** Frontal plane QRS-T angles between 100° and 180° indicate that the T waves in these cases are abnormal. This is usually true except in the neonate. In the normal newborn, the QRS vector is directed to the right and anterior. The T vector is oriented to the left and posterior creating a QRS-T angle between 90° and 160°. This wide angle is the result of right ventricular preponderance present in the normal neonate. T waves will be inverted in V_1 through V_4, and occasionally in V_5.

By the age of two years, however, the mean QRS vector is shifted to the left to about +60°. The T vector also moves to the left and posterior creating a QRS-T angle of 45° to 90°. There are no significant changes of the angle throughout adolescence and early adulthood.

The mean frontal QRS and T vectors in Example 3 have been located at 0° for the QRS and +30° for the T. The frontal plane QRS-T angle, therefore, is 30° which is normal (Figure 15 A).

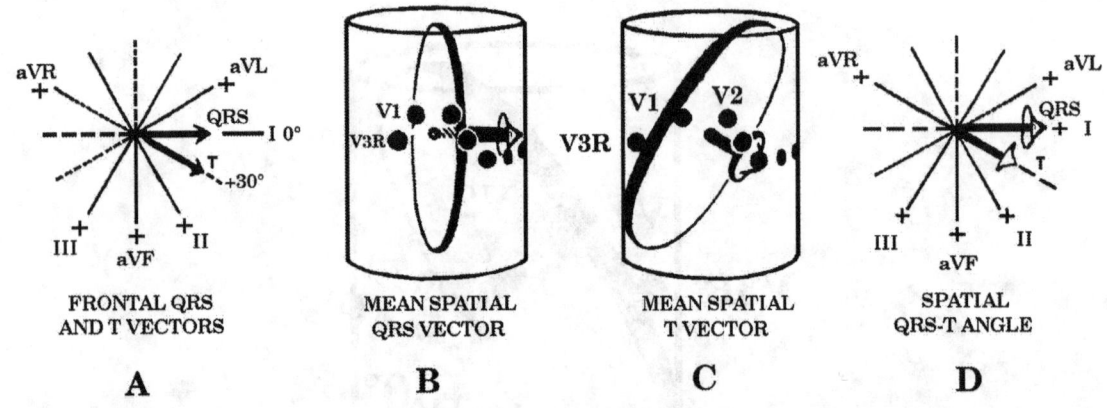

FRONTAL QRS AND T VECTORS   MEAN SPATIAL QRS VECTOR   MEAN SPATIAL T VECTOR   SPATIAL QRS-T ANGLE

A   B   C   D

Figure 15

12. Mean P Wave Vector

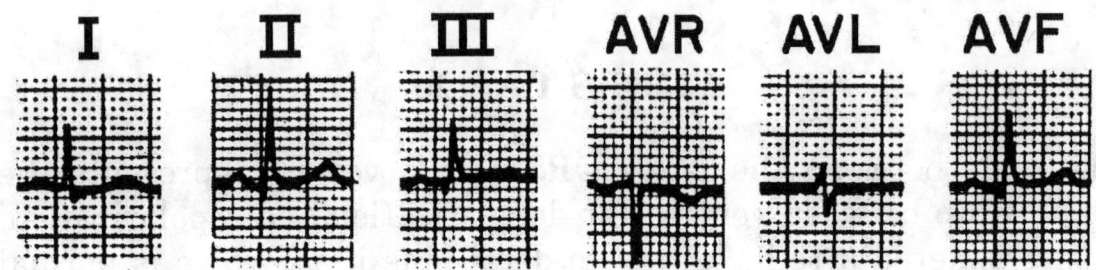

Example 5

Mean P vectors are determined in the same manner as the mean QRS and T vectors. In the six-lead electrocardiogram in Example 5, the P wave is positive in I and aVF. This indicates that the mean P vector is located in the left inferior quadrant. There is a flat (zero) P wave in aVL indicating that the mean P vector is perpendicular to the axis of aVL in the left inferior quadrant at +60° (Figure 16).

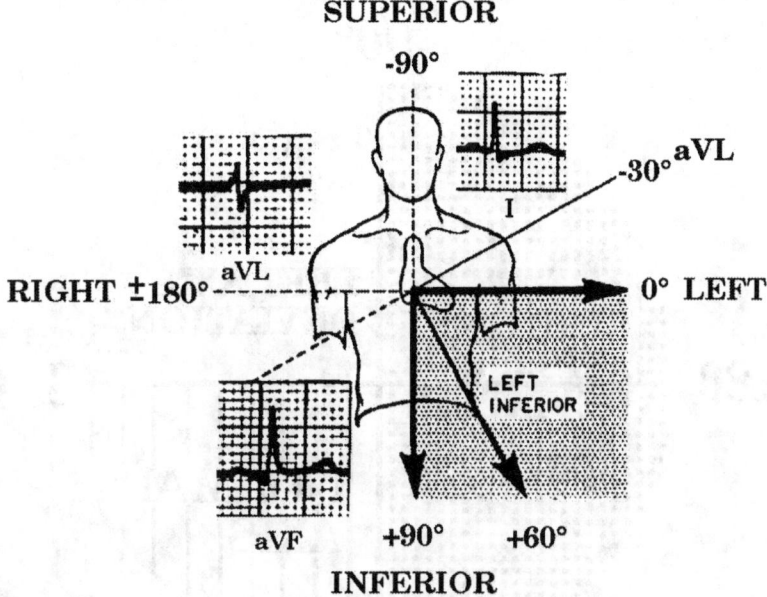

Figure 16

B. Axis Deviation

The orientation of ventricular electrical activity in the **frontal plane** is referred to as the axis and is expressed in degrees. The axis may be normal or there may be an abnormal deviation of the axis to the left or right. The location of the axis is determined by localizing the mean frontal QRS vector with which it is synonymous (Figure 17).

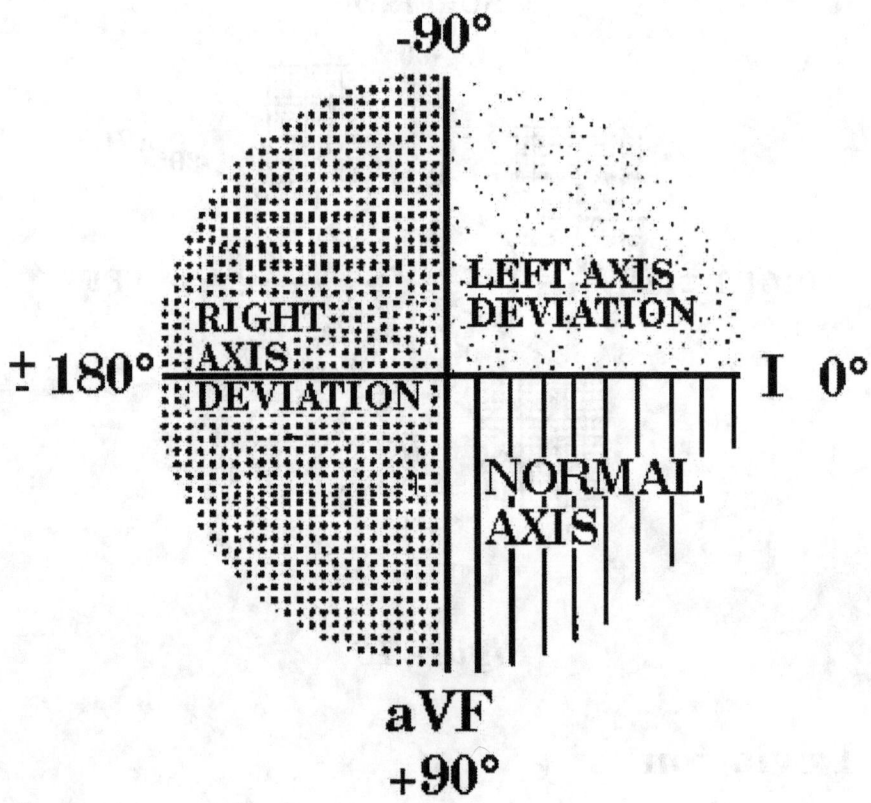

Figure 17

1. Left Axis Deviation

Left axis deviation (LAD) is present when the mean frontal QRS
vector lies between 0° and –30°. **Abnormal** LAD is present in adults
when the mean frontal QRS vector is located between –30° and –90°.
Abnormal LAD is present in children three months to sixteen years
of age when the mean frontal QRS vector lies superior to 0°. In infants
one week to three months of age, LAD is present when the mean
frontal QRS vector lies superior to +20°.

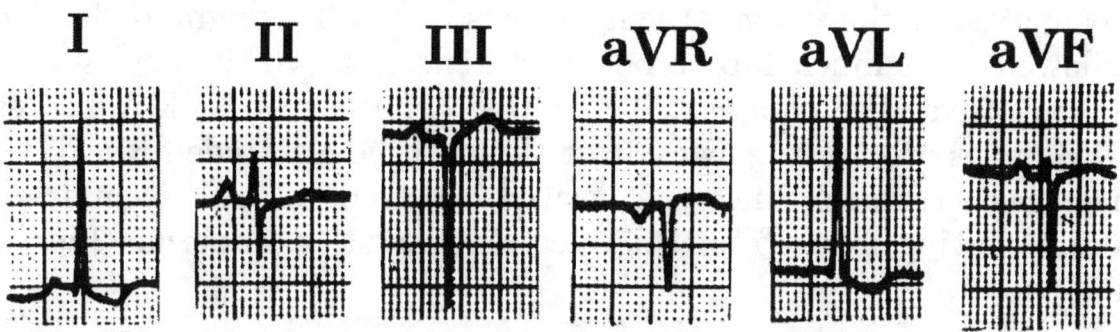

Example 6

In Example 6, there is a positive deflection, an R wave, in lead I. This indicates that the QRS vector is directed to the left toward the positive electrode of lead I (Figure 18A). A rightward QRS vector would require a negative deflection in lead I. However, the deflection in lead I is positive (Figure 18B).

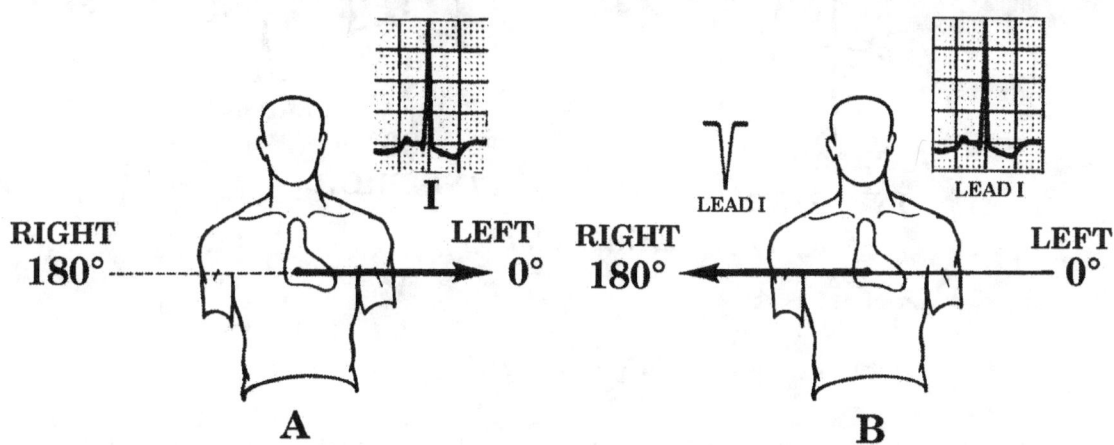

Figure 18

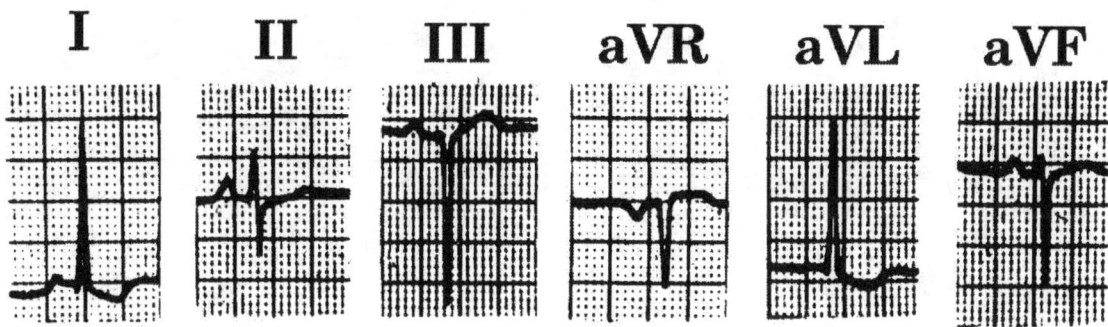

Lead aVF in the above tracing shows a small upright deflection followed by a much larger negative deflection which results in a net negative QRS complex. A negative deflection in aVF indicates QRS forces are moving away from the positive electrode toward the negative electrode. Since the positive electrode of aVF is inferior, forces moving away from this electrode must be superior (Figure 19A).

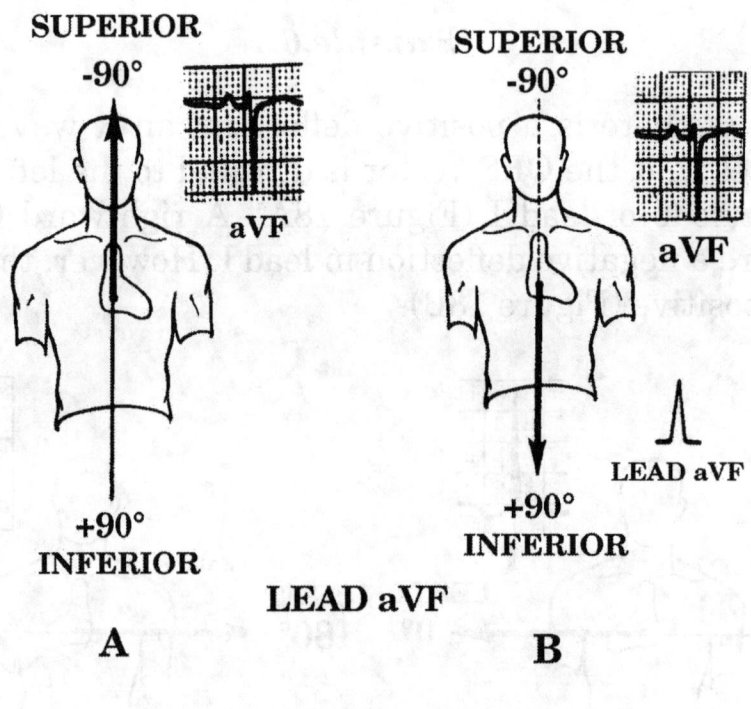

Figure 19

An inferior QRS vector demands a positive deflection in aVF. The QRS complex in aVF is negative indicating that the QRS vector has to be moving superiorly (Figure 19B).

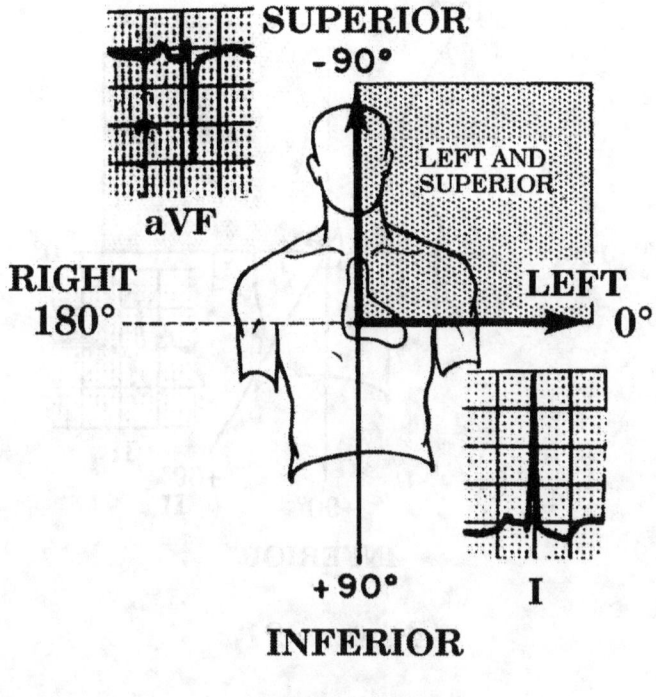

Figure 20

The positive QRS complex in lead I combined with the negative QRS complex in aVF localizes the mean QRS vector for this tracing to the *left superior* quadrant (Figure 20).

The mean QRS vector may be localized more precisely in degrees by the Perpendicular Rule of Spatial Analysis. This rule states that the mean QRS vector is perpendicular to the axis of the lead with the equiphasic complex. Lead II of this tracing displays an equiphasic QRS. The mean frontal QRS vector must be located in the left superior quadrant at –30° (Figure 21). This is consistent with left axis deviation but is not necessarily abnormal.

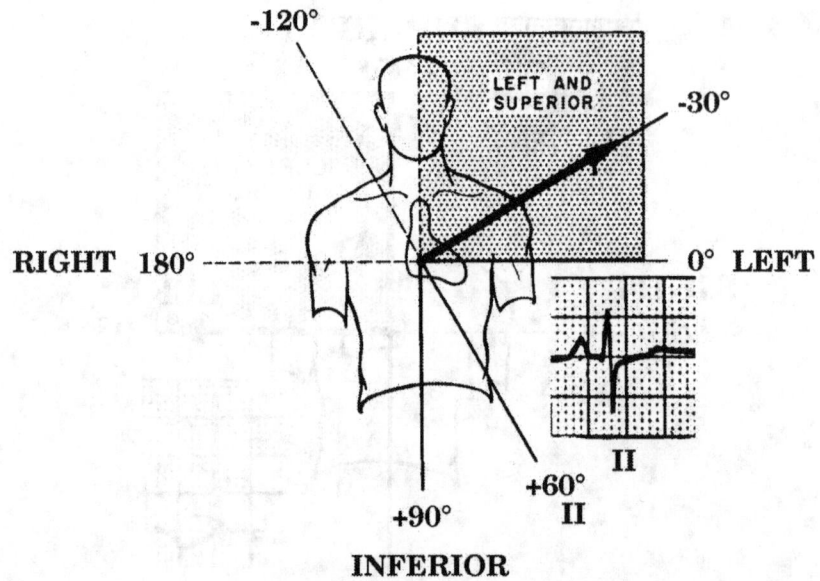

Figure 21

2. Right Axis Deviation

Right axis deviation (RAD) is present when the mean frontal QRS vector lies between +90° and –90° (Figure 18). ***Abnormal*** RAD is present in adults when the mean frontal QRS vector lies between +110° and –90°. A mean frontal QRS vector between +90° and +110° is frequently a normal variant.

Abnormal RAD is present in children three months to sixteen years of age when the mean frontal QRS vector exceeds +120°. ***Abnormal*** RAD is probably present in infants one to three months of age if the mean frontal QRS vector exceeds +140°.

The most frequent cause of RAD is right ventricular hypertrophy (See Chapter 4, page 133). It is also seen in emphysema and left posterior hemiblock.

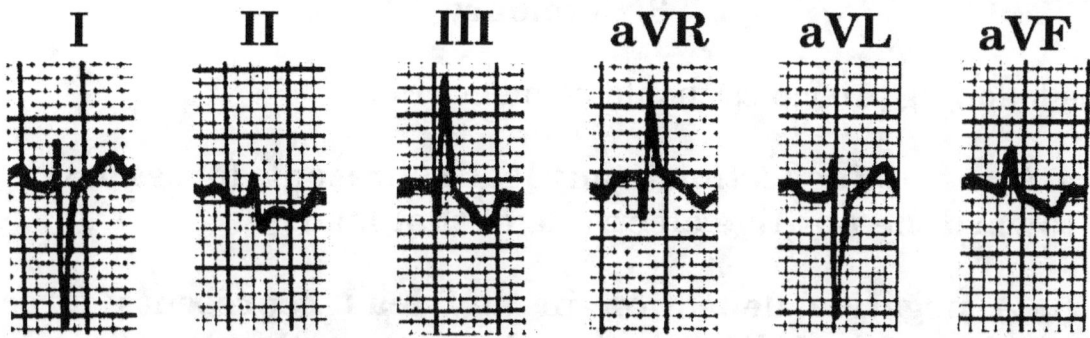

Example 7

In Example 7, the QRS is negative in I and positive in aVF. The mean frontal QRS vector is directed rightward and inferior. The equiphasic QRS complex in located in II. The mean frontal QRS vector has to be perpendicular to the axis of II in the right inferior quadrant at +150° (Figure 22). This is consistent with abnormal RAD.

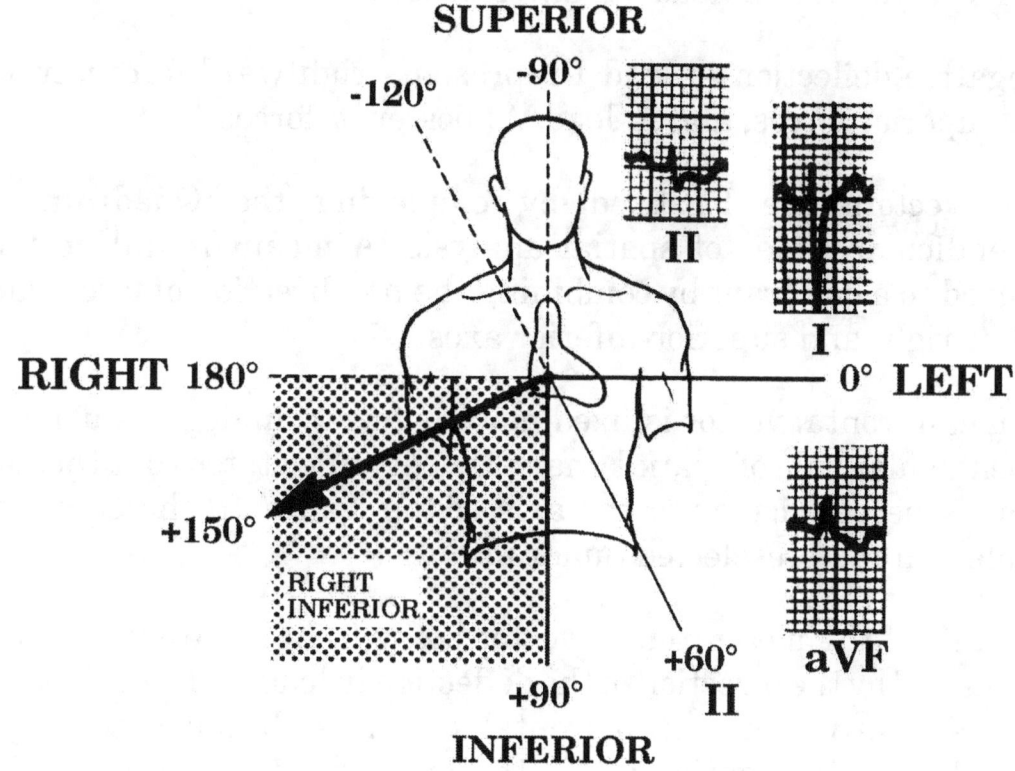

Figure 22

3. Summary

The following principles should be memorized:

a. A positive deflection in any lead represents forces moving toward the positive electrode of that lead.

b. A negative deflection in any lead represents forces moving toward the negative electrode of that lead.

c. The lead with the best left - right axis is I.

d. The lead with the best superior - inferior axis is aVF.

e. The lead with the best anterior - posterior axis is V_2 (V_1).

A positive deflection in lead I represents leftward forces, in lead aVF, inferior forces, and in lead V_2, anterior forces.

A negative deflection in lead I represents rightward forces, in lead aVF, superior forces, and in lead V_2, posterior forces.

Mean vectors are localized by employing the Quadrant and Perpendicular Rules of spatial analysis. A mean frontal vector is localized to a quadrant by combining the net direction of forces along the left-right and superior-inferior axes.

The mean frontal vector is fixed more precisely in degrees using the perpendicular rule of spatial analysis. This rule states that the mean vector is perpendicular to the axis of the lead with the equiphasic complex, in the preselected quadrant.

Any shift of the mean frontal vector away from the frontal plane is determined by the direction of the deflection in lead V_2. If the deflection in V_2 is positive, the mean frontal vector is shifted anteriorly. If the deflection is negative in V_2, the mean frontal vector is shifted posteriorly. The only exception to this is when there is marked right axis deviation of the mean frontal vector.

4. Normal Values

a. The normal mean QRS vector for adults lies between 0° and +90° with a posterior rotation of 10 to 40 degrees.

b. The normal mean T vector for adults lies between 0° and +90° with an anterior rotation of 10 to 40 degrees.

c. The normal mean frontal P vector for adults lies between +5° and +90°.

CHAPTER FOUR:
CHAMBER ENLARGEMENT AND HYPERTROPHY

Atrial Abnormalities

General Considerations

In normal sinus rhythm, the S-A node is the pacemaker of the heart. The wavefront of depolarization moves left and inferiorly from the S-A node to the A-V node. Depolarization spreads first to the right atrium, then to the left atrium and finally to the ventricles. Inter-atrial tracts speed up the atrial activation process (Figures 14, 15, Chapter 1, Conduction System).

The normal P wave is not easily defined and its normal range is wide. P waves are the result of forces generated by the right and left atria. Initial P forces are produced by depolarization of the right atrium, terminal P forces by depolarization of the left atrium and mid P forces by both. The right atrial P vector is directed anteriorly, the left atrial P vector posteriorly. P wave duration ranges from 0.07 sec. in infants to a maximum of 0.10 sec in adults. P wave amplitude normally does not exceed 2.0 mm (Figure 6). In normal sinus rhythm, neither atrium is electrically dominant. The normal mean P wave vector is oriented in the frontal plane between about + 5° to + 90° but is generally near +60° (Figure 1, Example 1).

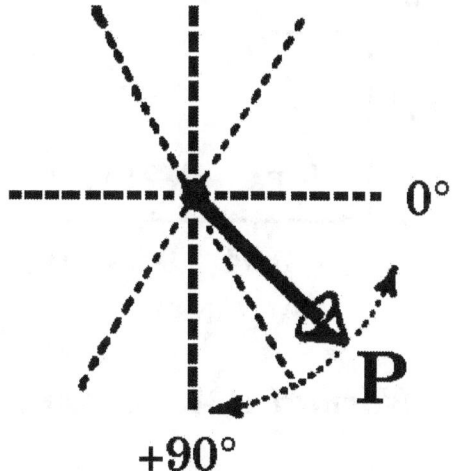

Figure 1: Range of Normal P Vector

1. Normal P Wave

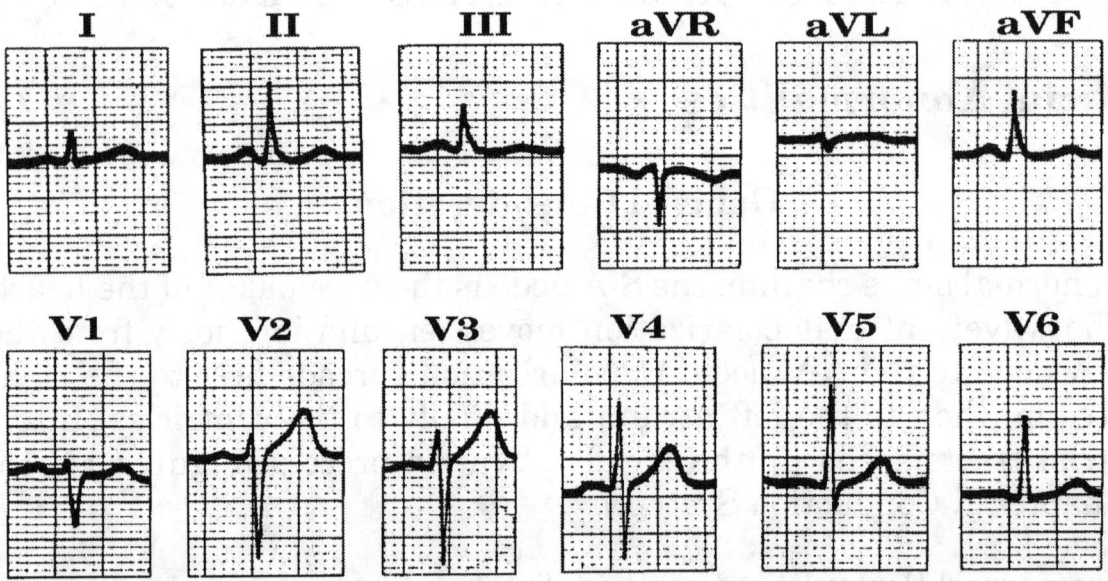

Example 1: Normal P Vector

In Example 1, the P is positive in I and aVF, and flat in aVL. This places the mean frontal P vector in the left inferior quadrant perpendicular to the axis of aVL, at +60°. The P is positive in V_2 indicating that the P vector is oriented anteriorly.

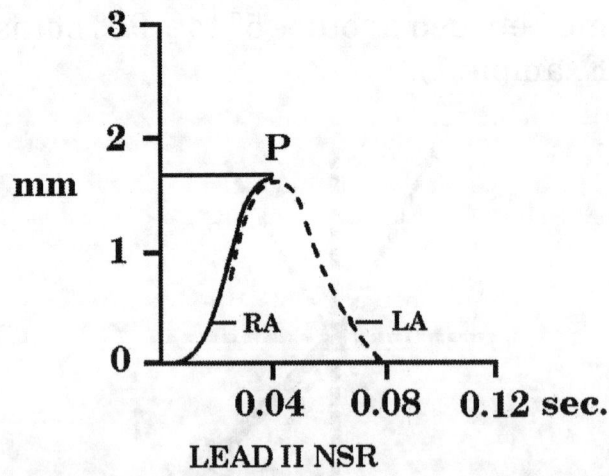

Figure 2: Normal P Wave Configuration

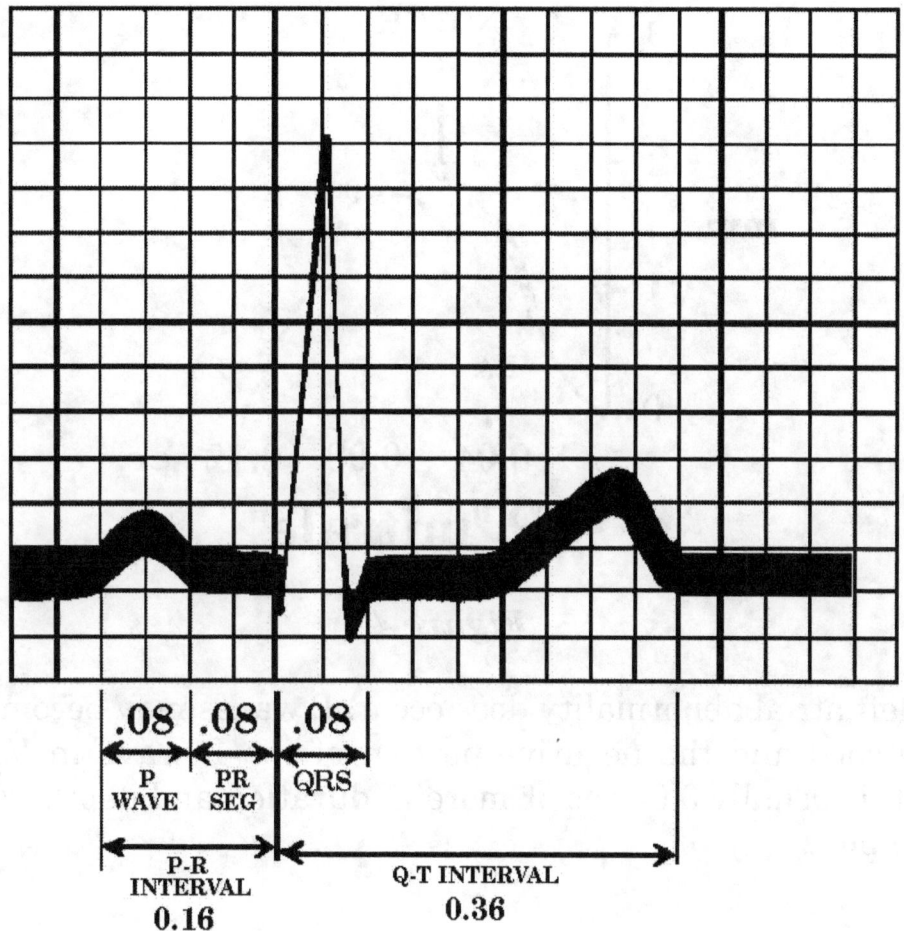

Figure 3

When both atria are normal, the ratio of the duration of the P wave to that of the PR segment is about one-to-one. In the above illustration, the duration of the P wave and PR segment each measure 0.08-sec, consistent with a one-to-one ratio. (Figure 3).

2. Left Atrial Abnormality

The term "left atrial abnormality" is used because it is not clear what causes the P wave abnormality. P waves may exhibit an abnormal configuration because of an intra-atrial conduction defect, and not because of left atrial enlargement. Left atrial abnormalities may be due to mitral stenosis, mitral insufficiency, systemic hypertension and left ventricular failure, just to mention a few. P waves may be notched in the absence of disease.

107

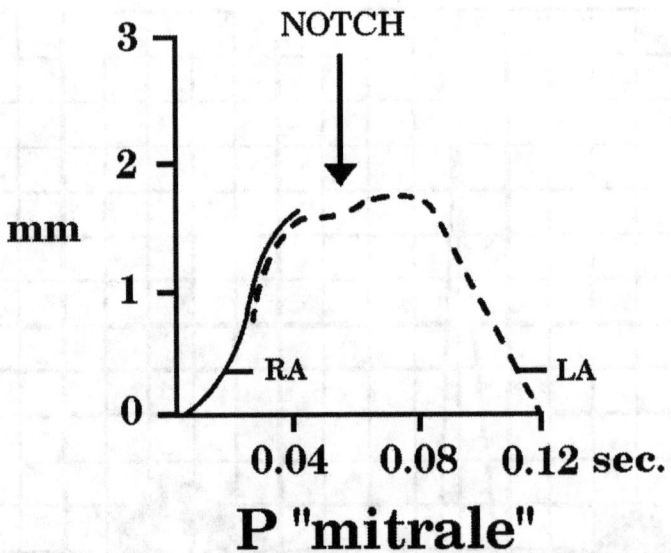

Figure 4

When left atrial abnormality does occur, P waves may become broad and notched and the negative portion of the P wave in V_1, when present, is usually 0.04 sec or more in duration and depth (Figure 4 & Example 2).

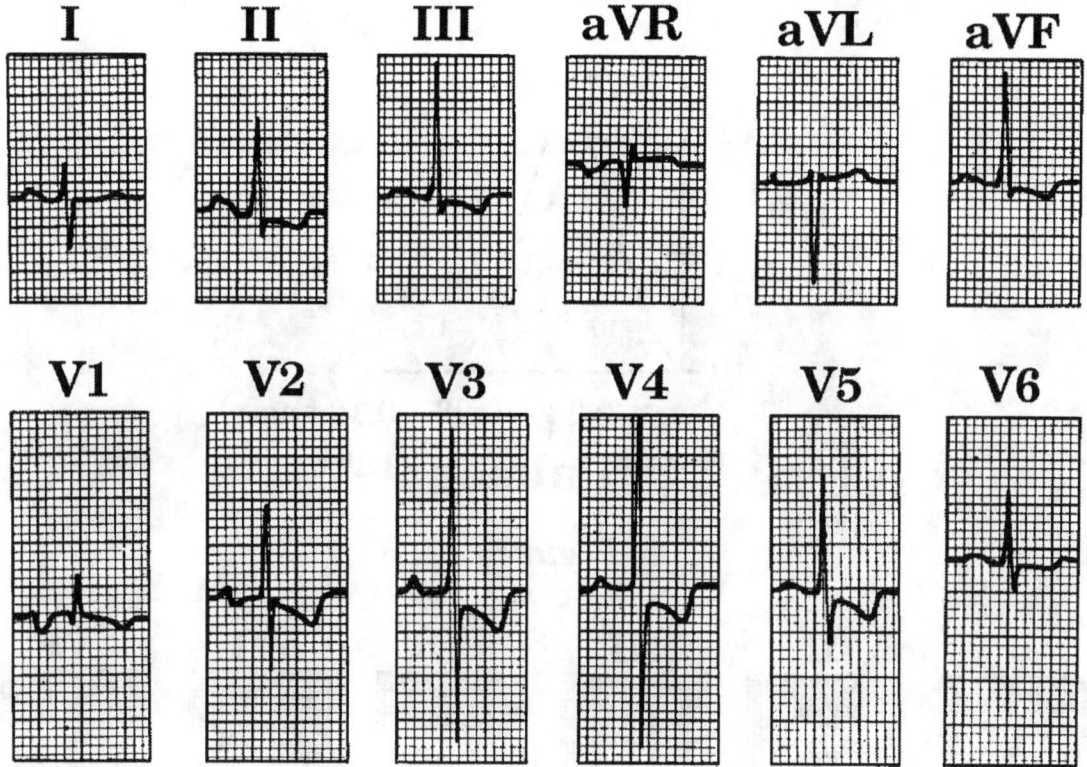

Example 2: Mitral Stenosis

In Example 2, P waves are broad and notched in leads II, III, and aVF. The P is positive in I and aVF and flat in a VL, which places the mean P vector in the left inferior quadrant perpendicular to the axis of a VL at +60°. Both the duration and depth of the negative P wave in V_1 exceed 0.04 sec. This is highly suggestive of left atrial enlargement. In this patient, the cause of the P wave abnormality is mitral stenosis.

3. Right Atrial Abnormality

The term "right atrial abnormality" is used because it is not certain what causes the P wave abnormality. Right atrial abnormality may be seen in patients with mitral stenosis, pulmonary emphysema, pulmonary valve stenosis and left ventricular failure, to mention a few. In cases of right atrial abnormality, the P wave becomes tall and peaked. Frequently, however, P waves are peaked in the absence of disease. (Figure 5).

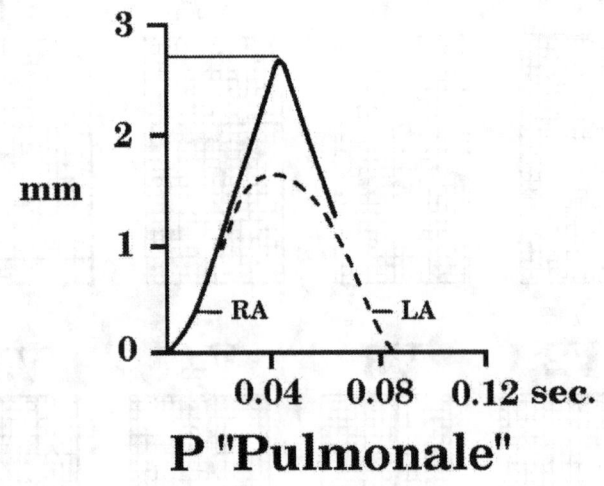

P "Pulmonale"

Figure 5

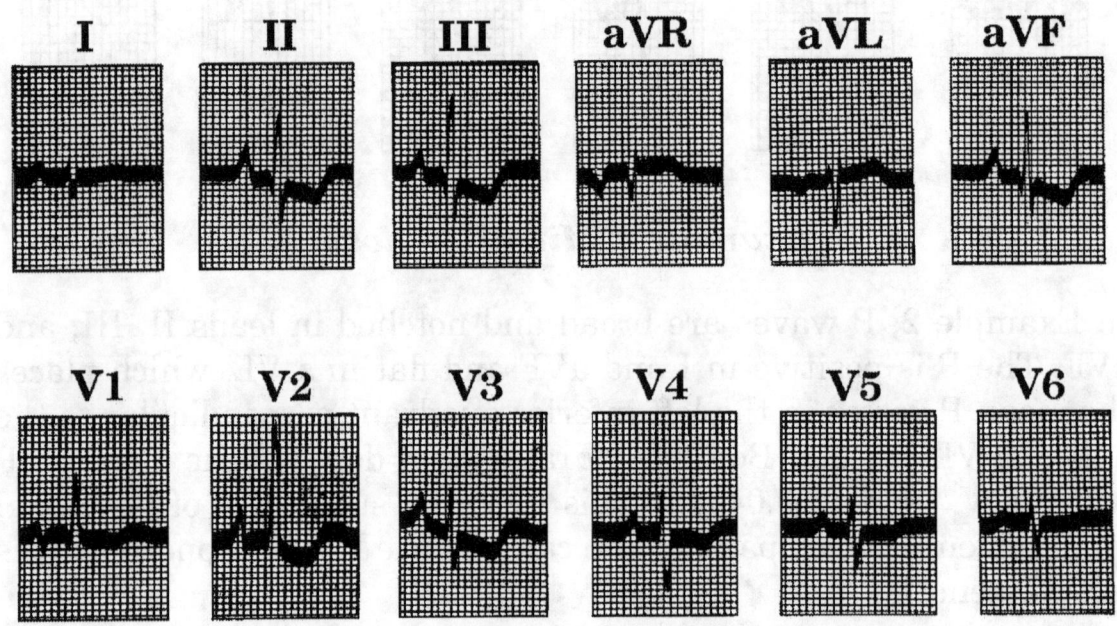

Example 3: Chronic Cor Pulmonale

In Example 3, note the peaked P waves in II, III, aVF, V_2 and V_3. A positive P in I and aVF indicates that the P vector is directed to the left and inferior. An equiphasic P in aVL places the P vector perpendicular to aVL at + 60°. Right ventricular hypertrophy is present (Please refer to Example 10 on page 138).

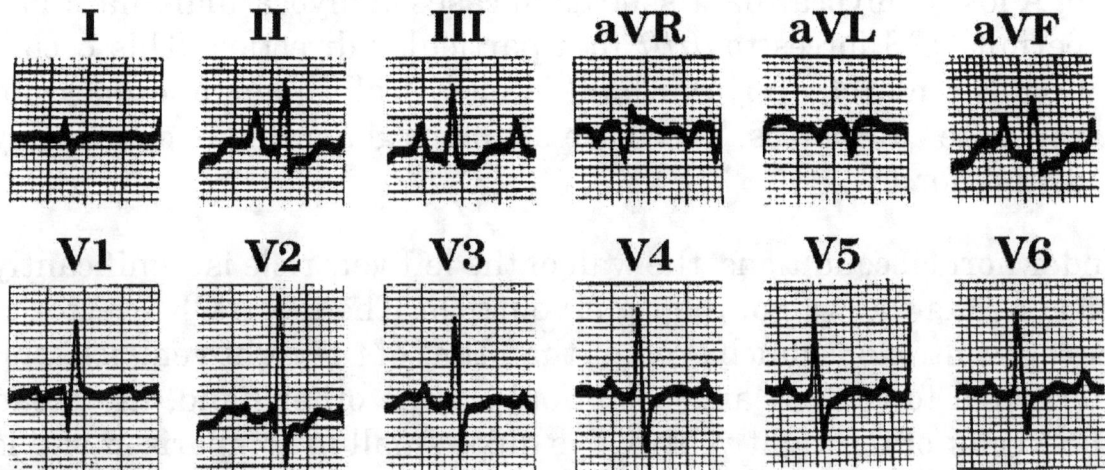

Example 4: P Pulmonale

The P waves are tall and peaked in Example 4, also. The flat P in I and positive P in aVF indicates that the P vector is at + 90°

4. Bi-Atrial Abnormality

Right and left atrial abnormalities may occur in the same patient. In these cases, there is an increase in amplitude and duration of the P wave. Long-standing mitral stenosis often produces left atrial abnormality and when sufficient pulmonary hypertension develops, right atrial abnormality will occur.

Ventricular Hypertrophy

General Considerations

Shifting of the mean QRS vector may be the result of two entirely different mechanisms:

a. An abnormally large myocardial mass, as seen in cases of ventricular hypertrophy, *pulls* QRS forces to a particular direction. This occurs because the hypertrophied muscle generates more voltage.

111

b. A loss of myocardial tissue as in cases of myocardial infarction, permit QRS forces to *drift* to a particular direction. This occurs because infarcted myocardium is electrically inert and does not generate QRS forces. There are no forces generated to oppose any drift of forces.

Under normal conditions, the wall of the left ventricle is significantly thicker than the wall of the right ventricle. The mean QRS vector of the left ventricle is much greater than that of the right ventricle and is oriented to the left and posterior. On the other hand, the mean QRS vector of the right ventricle is quite small in comparison and is directed anteriorly. (Figures 6 & 7).

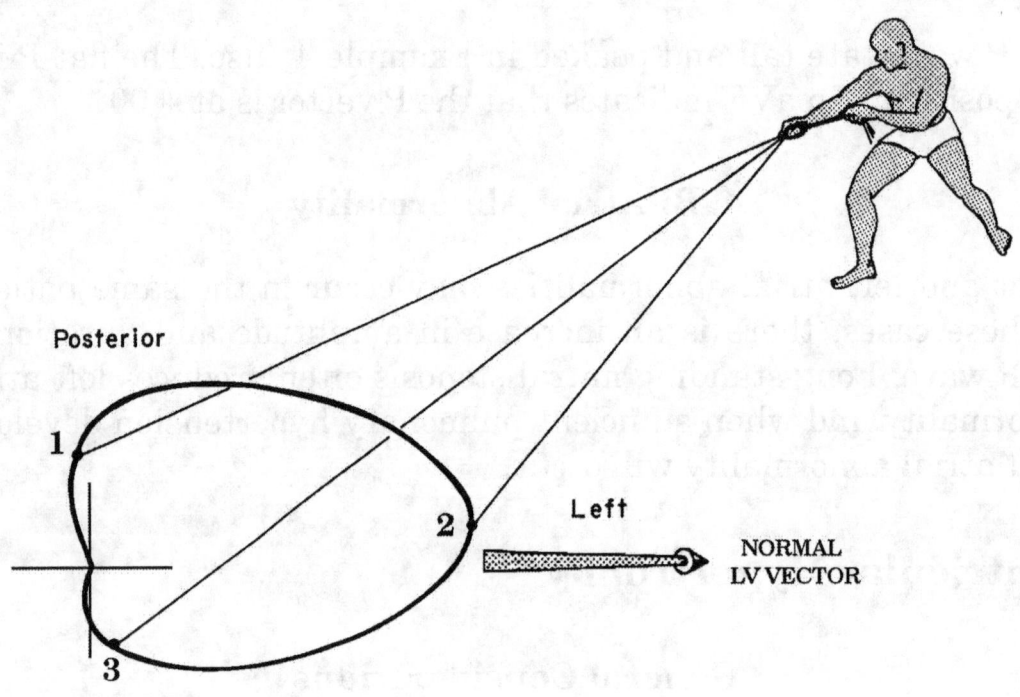

Figure 6

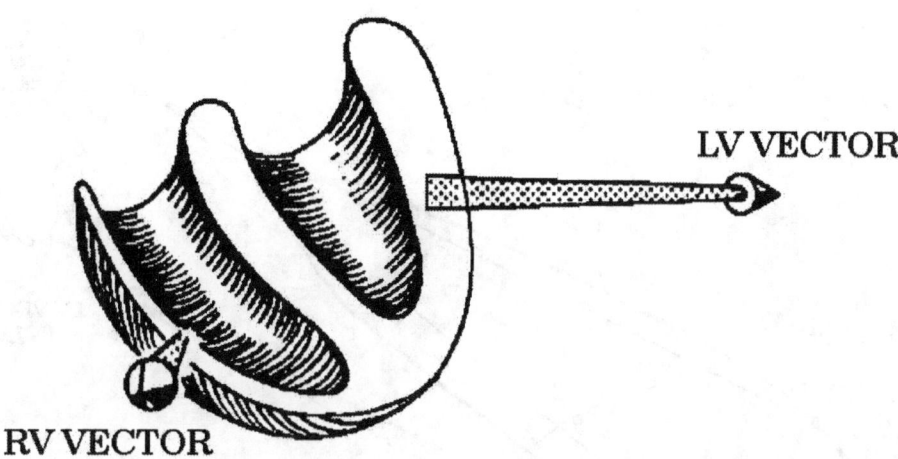

Figure 7: Normal Mean Ventricular Vectors

Left Ventricular Hypertrophy

The left ventricle may increase in size for a variety of reasons. These include, but are not limited to, high blood pressure and aortic valve stenosis. When left ventricular hypertrophy does occur, the mean QRS vector tends to be increased in magnitude and deviated superiorly. (Figures 8 & 9).

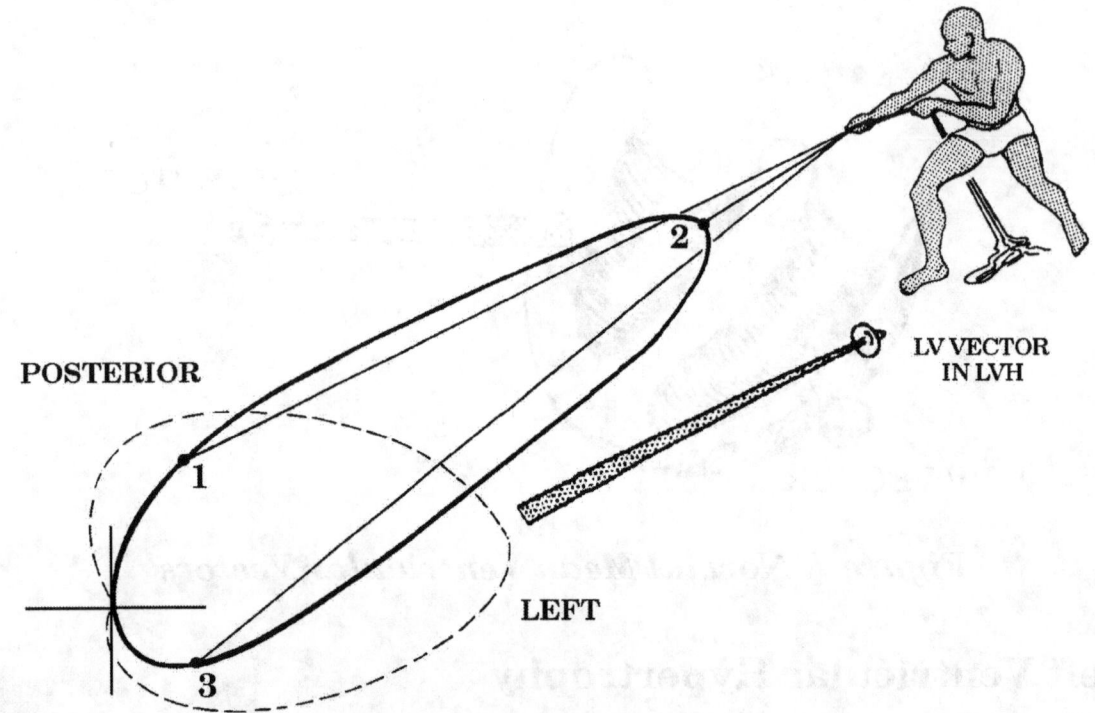

POSTERIOR

1

2

3

LEFT

LV VECTOR
IN LVH

Figure 8

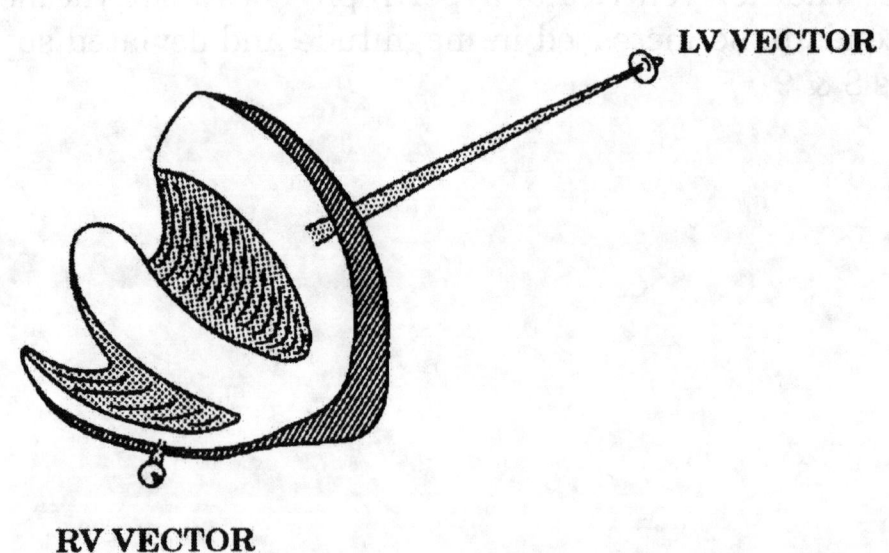

LV VECTOR

RV VECTOR

Figure 9

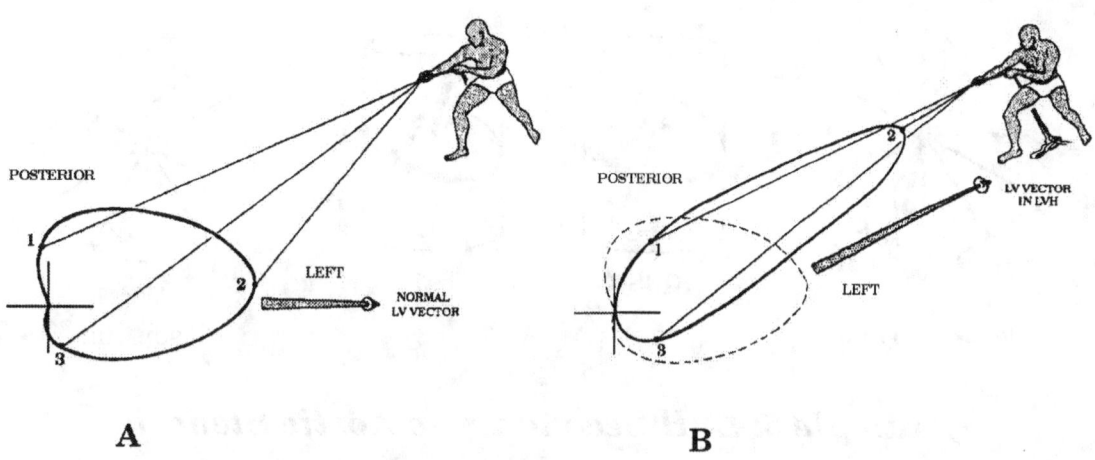

POSTERIOR

1

2

LEFT

NORMAL
LV VECTOR

A

POSTERIOR

2

1

3

LEFT

LV VECTOR
IN LVH

B

Figure 10

The mean QRS vector points toward the predominant muscle mass. Consequently, in left ventricular hypertrophy (LVH), the mean QRS vector is larger than normal and directed more to the left, superiorly, and posteriorly (Figures 8, 9 & 10). However, LVH alone usually will not shift the mean QRS vector beyond minus 30°. In fact, in some cases there may be no shift at all. Left axis deviation alone is not diagnostic of LVH but it supports the diagnosis when voltage criteria are present.

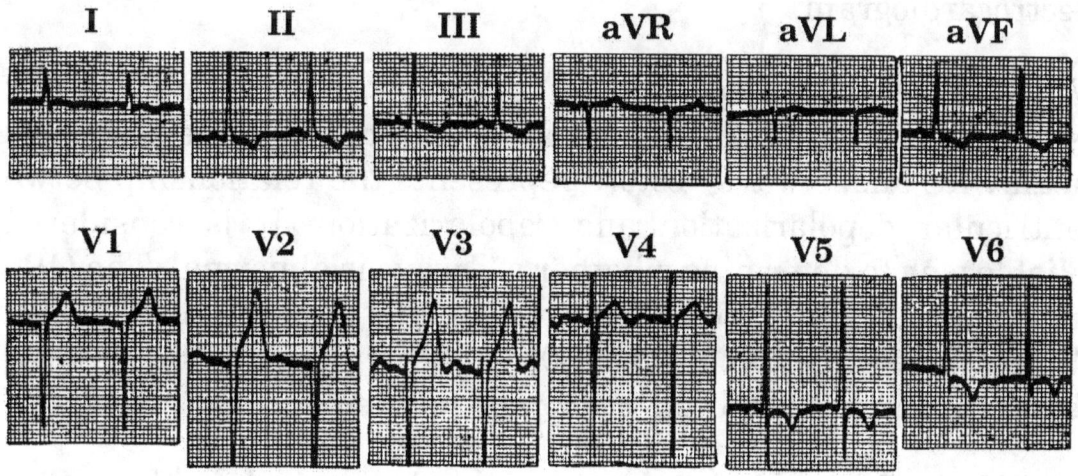

I   II   III   aVR   aVL   aVF

V1   V2   V3   V4   V5   V6

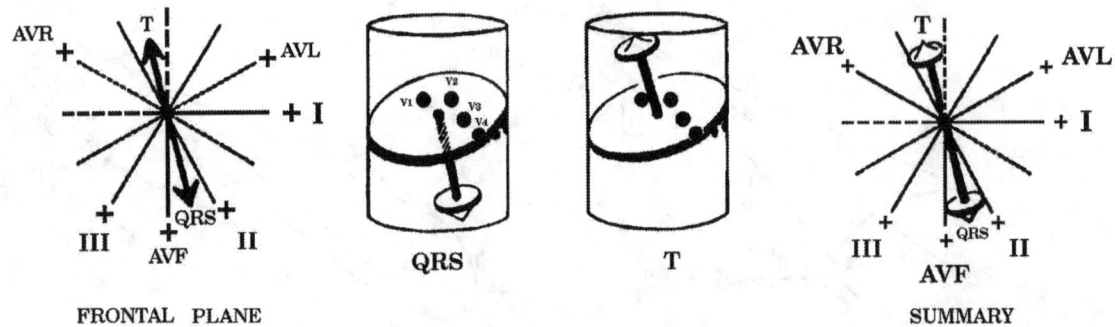

FRONTAL PLANE   QRS   T   SUMMARY

Example 5: LVH Secondary to Aortic Stenosis

The tracing in Example 5 is an adult with aortic stenosis. The QRS complexes are positive in I and aVF placing the mean QRS vector to the left and inferiorly. No lead has an equiphasic complex, but II and aVF have R waves that are about equal in amplitude. This places the mean QRS vector between the positive axes of II and aVF, at +75°. The T is negative in I and aVF indicating that the T vector is oriented to the right and superiorly. The area under the T waves in I and aVL are about equal but negative in I and positive in aVL. This places the mean T wave vector between the perpendiculars of the axes of I and aVL at minus 105°. The QRS-T angle is abnormally wide at 180° indicating that T wave abnormalities are present in this electrocardiogram.

When ST segment displacement is present, the mean ST vector is usually almost parallel to the mean T vector. The angle between the mean QRS and T wave vectors represents the relationship between ventricular depolarization and repolarization. It is considered a reliable way to determine whether a T wave is abnormal. The QRS-T angle rarely exceeds 45° in the frontal plane or 60° in the transverse plane. When QRS-T angles of 100° to 180° are found, the T vector is abnormal for that QRS vector, *except in the neonate*.

In LVH, the T vector is frequently oriented to the right and superiorly. This coupled with a QRS vector that is displaced to the left and inferiorly results in an abnormally wide QRS-T angle (Example 5).

The EKG in LVH may be normal or show specific or nonspecific abnormalities. Criteria for the diagnosis of LVH are valid only when the QRS interval is less than 0.12 sec. The diagnosis of LVH is based primarily on increased voltage of those QRS deflections representing left ventricular forces.

Voltage Criteria for LVH

a. The sum of the R wave in V_5 or V_6 and of the S wave in V_1 or V_2 greater than 35 mm

b. The sum of the highest R wave and deepest S wave in the precordial leads greater than 40 mm

c. The amplitude of the R wave in aVL greater than 11 mm

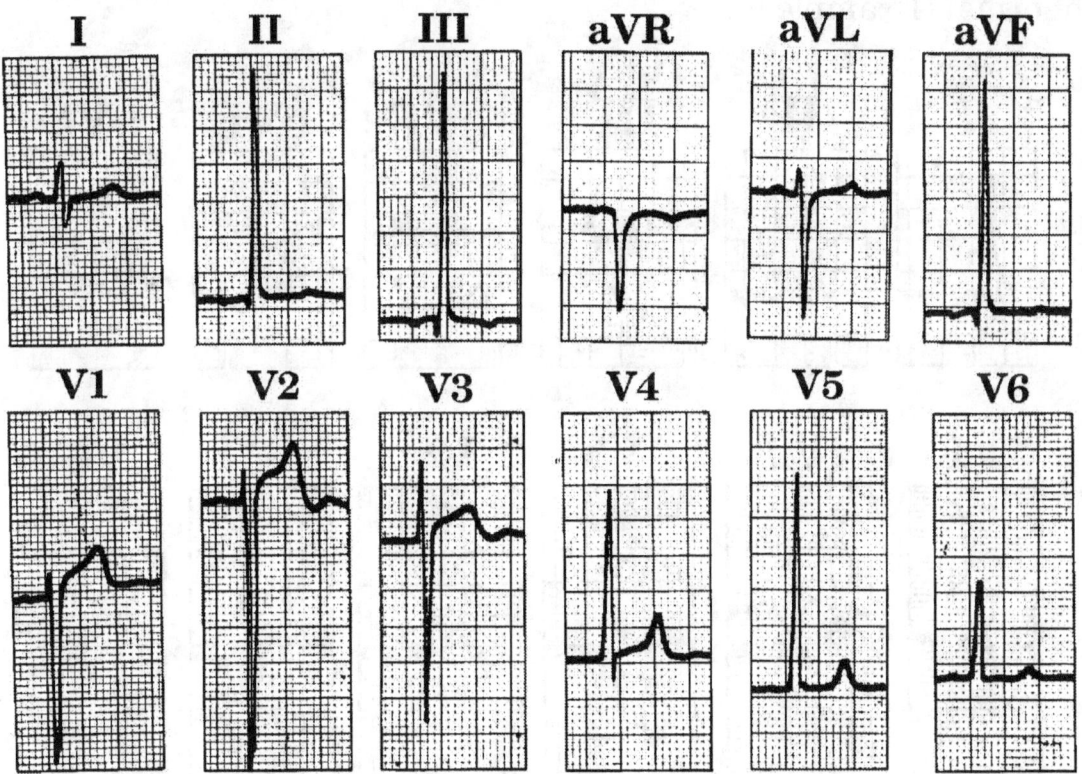

Example 6: LVH Secondary to Aortic Insufficiency

The tracing in Example 6 is an adult with an incompetent aortic valve. The R wave in V_5 measures about 30 mm and the S wave in V_2 measures around 37 mm. These values exceed the normal and are consistent with LVH.

117

The QRS complex in lead I is biphasic but slightly more positive than negative. This means QRS forces are moving to the left. The QRS complex in a VF is positive and measures about 29 mm. This places the mean QRS vector in the left inferior quadrant. The R wave in lead II also measures about 29 mm, the same as a VF. The mean QRS vector for this tracing must lie between +60° and +90°, or +75°. Since the QRS in V_2 is overall negative, the mean frontal QRS vector is shifted posteriorly.

Low amplitude T waves are present in the limb leads and there is terminal T wave inversion in V_1, V2 and V_3. The T in I is positive and flat (zero, the same as being equiphasic) in aVF indicating that the mean frontal T vector is at 0°. The frontal QRS-T angle is 75°, which is wider than normal. The T waves in this tracing, therefore, are abnormal (Example 6).

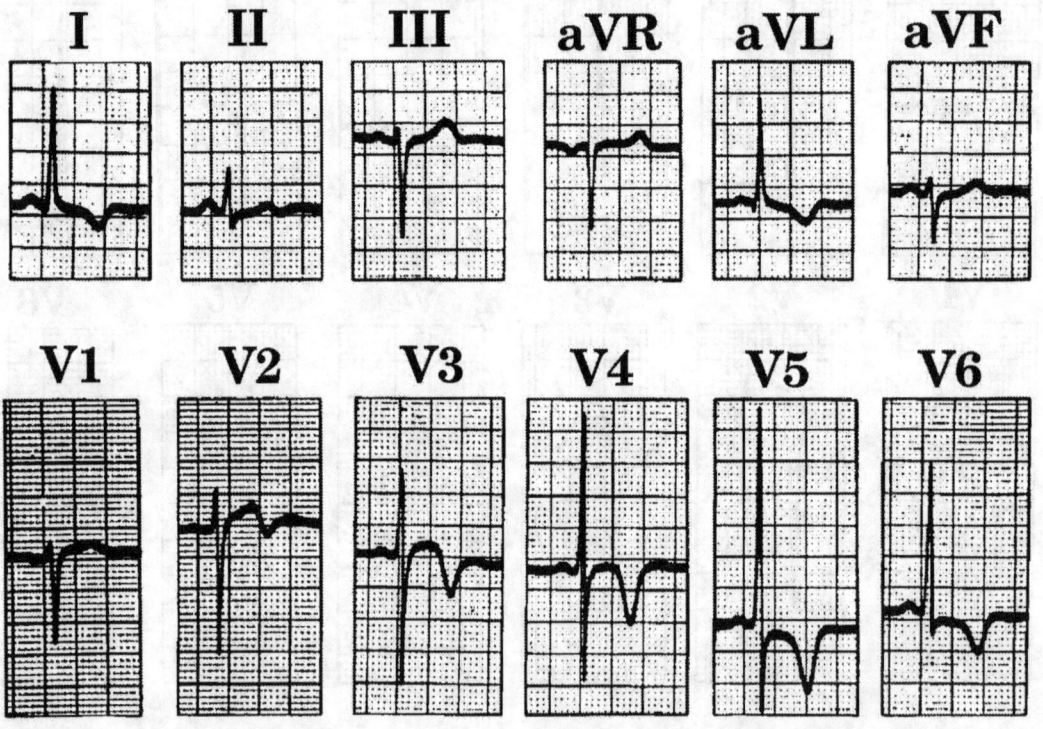

Example 7: LVH Secondary to Hypertension

The tracing in Example 7 is an adult with hypertensive heart disease. The R wave in aVL measures about 18 mm. The R wave in V_5 is around 37 mm and the S wave in V_1 is 12 mm. These values fulfill voltage criteria for LVH.

The QRS complex is positive in I and negative in aVF. This places the mean frontal QRS vector in the left superior quadrant. The QRS complexes in I and aVL are about equal in amplitude placing the mean QRS vector between 0° and -30° at -15°.

The T in I is negative and positive in aVF indicating that the mean frontal T vector is located in the right inferior quadrant. The T wave in II appears equiphasic placing the mean frontal T vector at about +150°. The frontal QRS-T angle of 165° is wider than normal indicating that the T waves in this tracing are abnormal.

Right Ventricular Hypertrophy

The right ventricle may increase in size for a variety of reasons. These include, but are not limited to, certain congenital heart abnormalities and pulmonic valve stenosis. When right ventricular hypertrophy does occur, the mean QRS vector increases in magnitude and tends to be redirected to the right and anterior (Figure 12).

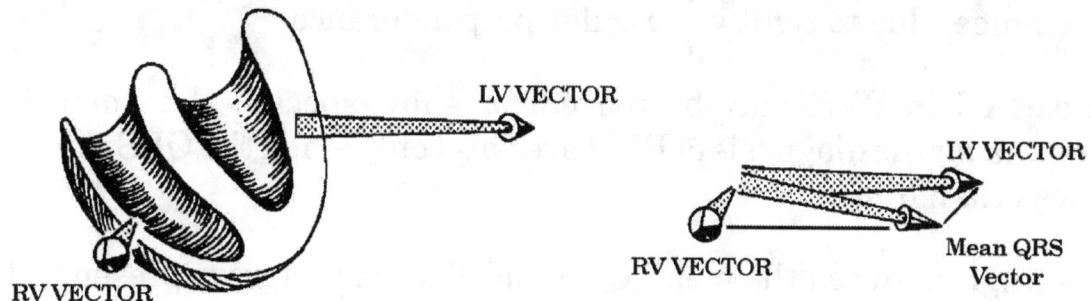

LV VECTOR

RV VECTOR

LV VECTOR

RV VECTOR

Mean QRS Vector

Figure 11: Normal Mean Ventricular Vectors

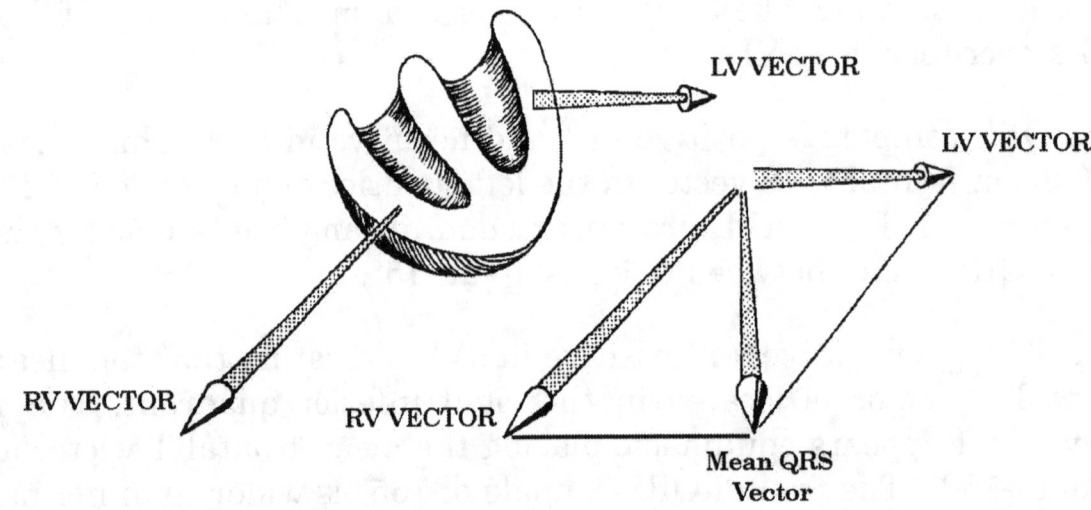

Figure 12: RVH

In the normal adult, the major ventricular mass is the left ventricle. The normal mean QRS vector is located in the left inferior quadrant between 0° and +90°. To overcome left ventricular predominance so that the specific EKG abnormalities of right ventricular hypertrophy (RVH) become apparent, the right ventricle must enlarge significantly. Even then, diagnostic abnormalities may not appear. The electrocardiograms of normal infants and children, on the other hand, are characterized by changes due to right ventricular preponderance.

The EKG in RVH may be normal or show specific abnormalities. Criteria for the diagnosis of RVH are valid only when the QRS interval is less than 0.12 sec.

The angle between the mean QRS and T wave vectors represents the relationship between ventricular depolarization and repolarization. It is considered a reliable way to determine whether a T wave is abnormal. This angle rarely exceeds 45° in the frontal plane or 60° in the transverse plane. When QRS-T angles of 100° to 180° are present, T waves in that tracing are abnormal, ***except in the neonate***.

Three types of RVH are distinguishable using spatial analysis. These are types A and B in which right ventricular QRS forces are directed anteriorly and type C where right ventricular forces are directed posteriorly. Type A is considered the most typical type of RVH and is seen most commonly in patients with pulmonic stenosis and tetralogy of Fallot.

Criteria for RVH

a. High voltage of the QRS deflections representing right ventricular forces (V_3R, V_1, V_2)

b. Initial anterior QRS forces prolonged to 0.04 sec or more, i.e., a wide R wave in V_1 or V_2

b. Abnormal right axis deviation

c. T vector is directed to the left

d. Abnormally wide QRS-T angle

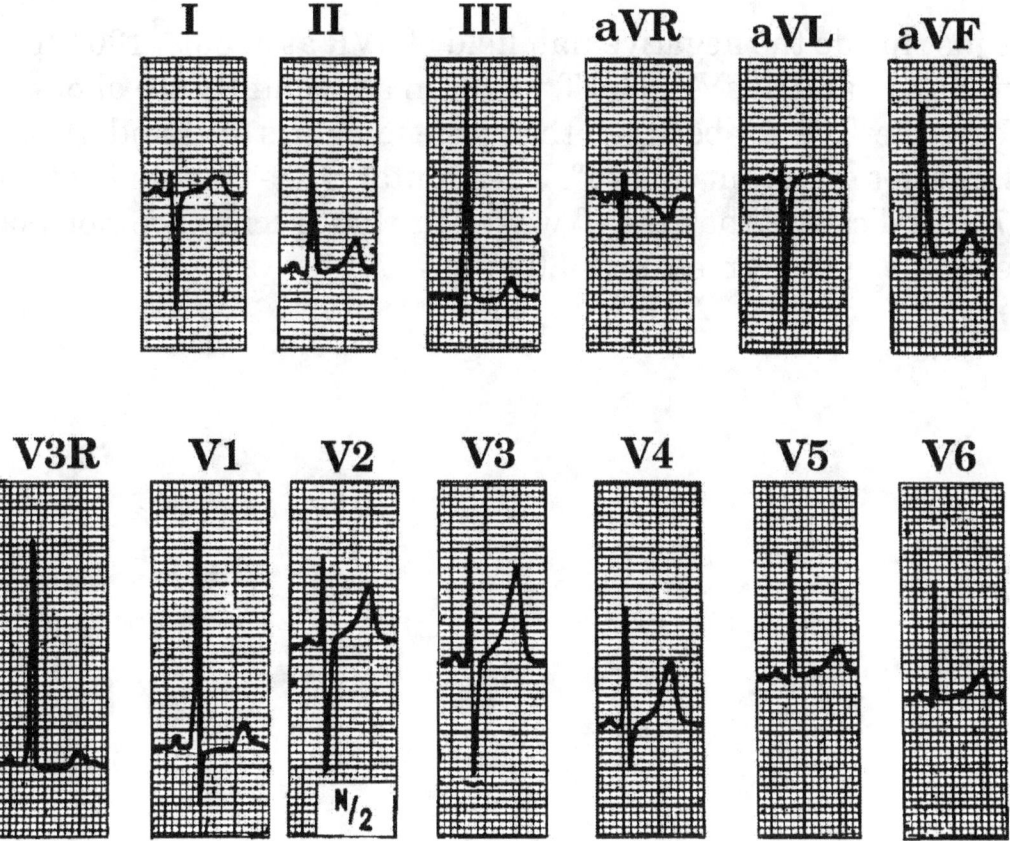

Example 8: Tetralogy of Fallot

Example 8 is a tracing of right ventricular hypertrophy in a 7-year-old with tetralogy of Fallot. This congenital cyanotic heart disease is characterized by

a. Pulmonary stenosis

b. Interventricular septal defect

c. Dextroposition of the aorta which straddles the ventricular septal defect

d. Right ventricular hypertrophy

The QRS deflection is negative in I and positive in aVF. The QRS is almost equiphasic in aVR, but slightly more negative than positive. The mean frontal QRS vector must be nearly perpendicular to the axis of aVR in the right inferior quadrant. This is confirmed by the R wave in III, which measures 30 mm.

It is just inside the negative half field of aVR at about +130°. The T is positive in I and in aVF. The T waves in aVL and III are about equal, placing the T vector between the perpendiculars of II and aVR in the left inferior quadrant at +45°. The frontal QRS-T angle is abnormal at 75°, indicating that the T waves in this tracing are not normal. The R wave in V_3R measures 30 mm.

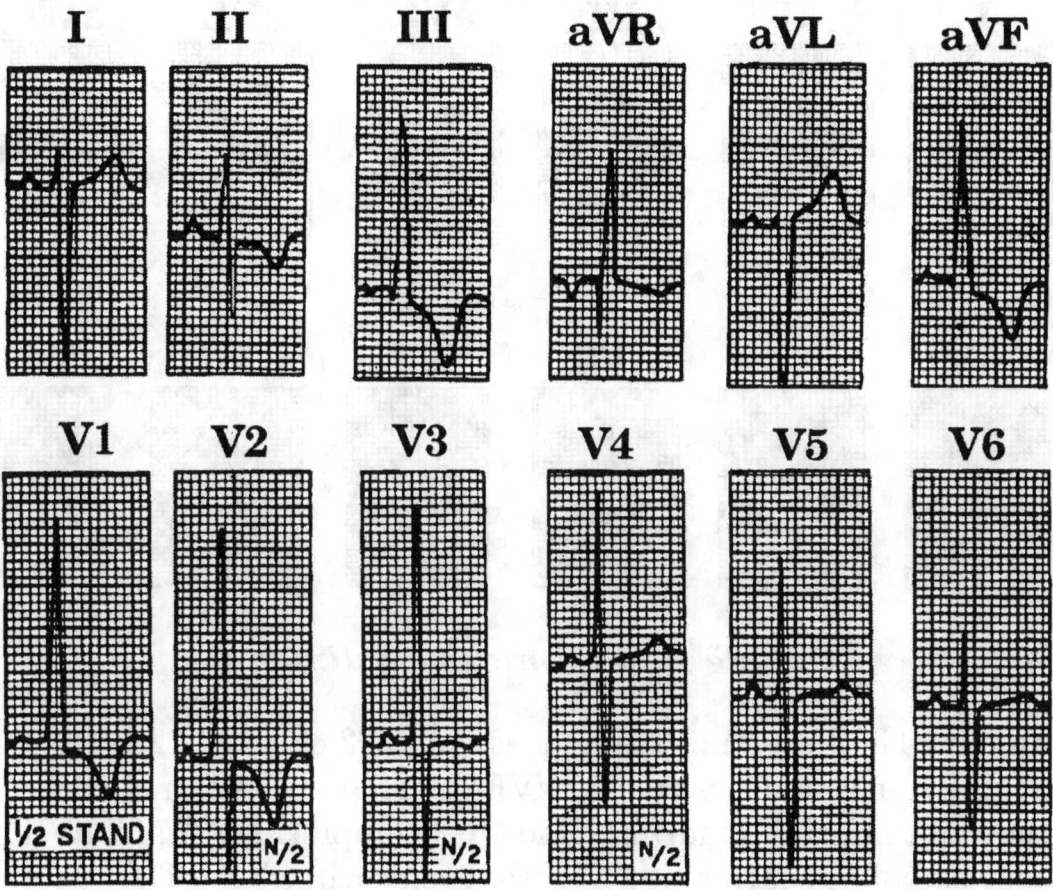

| I | II | III | aVR | aVL | aVF |

| V1 | V2 | V3 | V4 | V5 | V6 |

½ STAND N/2 N/2 N/2

Example 9: Pulmonary Valve Stenosis

The tracing in Example 9 is type A right ventricular hypertrophy in a young adult with pulmonic valvular stenosis. The QRS complex is negative in I and positive in aVF. This indicates that the mean frontal QRS vector is located in the right inferior quadrant. The QRS complex in II is equiphasic. The mean frontal QRS vector must be perpendicular to the axis of II in the right inferior quadrant at +150°. The T is positive in I, negative in aVF, and barely negative in aVR. Therefore, T wave forces are directed left and superior and close to being perpendicular to the axis of aVR just inside its negative half field at about -50°. The frontal QRS-T angle is abnormally wide at 200°. The inverted T waves in II, III, aVF, V_1, V_2 and V_3 are abnormal. The R wave in V_1 is one half standard measuring 50 mm.

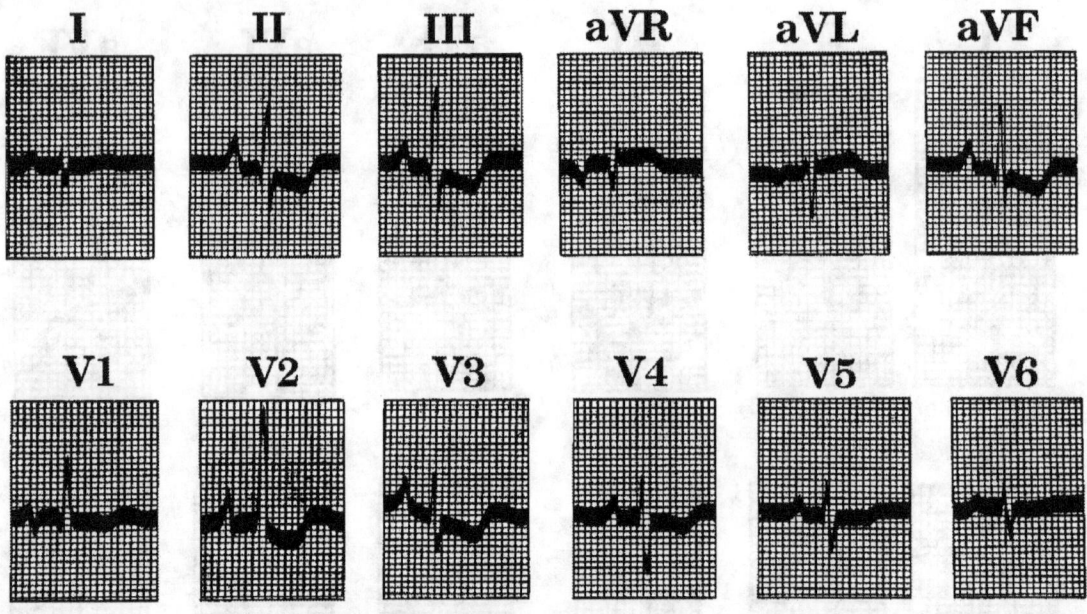

Example 10: Chronic Cor Pulmonale

Example 10 is the same tracing as Example 3 on page 124. The QRS is negative in I and positive in aVF, placing the mean QRS vector in the right inferior quadrant. The QRS complexes in I and aVR are almost equally negative but slightly greater in aVR. This puts the mean QRS vector at about +115°, consistent with right axis deviation. Initial anterior QRS forces (R wave in V_2) are very pronounced and exceed .04 sec in duration. The T wave is barely positive in I and decidedly negative in aVF. This places the T vector in the left superior quadrant. The T wave in aVR is equiphasic placing the mean T vector at −60°. The QRS-T angle is, therefore, 175°, which is abnormal. The criteria for RVH have been satisfied.

CHAPTER FIVE:
CONDUCTION DEFECTS

A. Ventricular Conduction System

1. Anatomic Considerations

A schematic diagram of the distribution of the ventricular conduction system is illustrated in Figure 1. The right bundle branch (RBB) and left bundle branch (LBB) with its two divisions are shown. The two bundle branches course down each side of the septum. The right bundle branch descends as a thin solitary cord until it reaches the apex of the right ventricle before it fans out to supply fibers to the right ventricle. The left bundle branch divides early into an anterior and posterior division and arborizes to supply fibers to the left ventricular side of the septum and the free wall of the left ventricle.

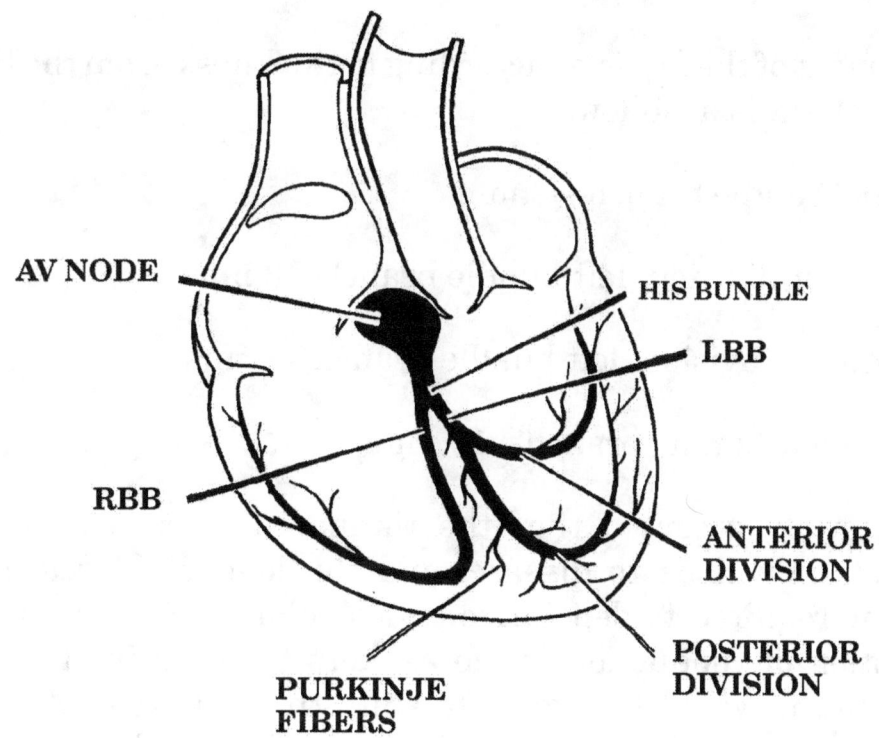

Figure 1

In the adult, the right bundle branch is a cylindrical fascicle with a diameter of 1 to 2 mm and an average length of 45 to 50 mm. It is a continuation of the bundle of His.

The left bundle branch splits off almost perpendicularly from the bundle of His and divides quickly into its anterior and posterior divisions. Anatomically the left and right bundle branches form a trifascicular ventricular conduction system.

Sorting of the fascicles according to length, from longest to shortest, is as follows:

    a. Right bundle branch—45 to 50 mm

    b. Anterior division, left bundle branch—25 mm

    c. Posterior division, left bundle branch—20 mm

    d. Main left bundle branch—10 mm

Arrangement of the fascicles according to thickness, from the thinnest to the thickest, is as follows:

    a. Right bundle branch—2 mm

    b. Anterior division, left bundle branch—3 mm

    c. Posterior division, left bundle branch—6 mm

    d. Main left bundle branch—10 mm

It is generally accepted that the thinner a fascicle is, the more vulnerable it will be to disease, and the longer a fascicle is, the more time required to depolarize and repolarize. As a consequence, interruption of conduction is more likely to occur in the thinnest fascicle and it would be logical to state that an order of anatomic vulnerability could be anticipated. The right bundle branch being the thinnest structure would be the most vulnerable, followed by the anterior division, the posterior division, and, finally, the main left bundle branch in order of decreasing vulnerability.

The clinical incidence of these blocks in order of decreasing frequency is:

a. Left anterior hemiblock,

b. Right bundle branch block

c. Left bundle branch block

d. Left posterior hemiblock

2. Electrophysiologic Considerations

Figure 2 is a schematic representation of the trifascicular anatomy of the ventricular conduction system. The impulse from the S-A node is received by the A-V node and moves to the bundle of His. It is delivered to the ventricular myocardium by way of the three fascicles labeled **"1, 2, 3"** in Figure 2. The earliest contact between conducting tissue and myocardium occurs at the mid-left septal endocardium. When septal activation is completed, the wavefront then moves from apex to base to finish the process. The subsequent ventricular contraction will follow the same sequence and cause the ventricles to contract from apex to base. In the presence of bundle branch block, however, the normal activation sequence is abnormal resulting in an abnormal contraction and could have significant hemodynamic consequences.

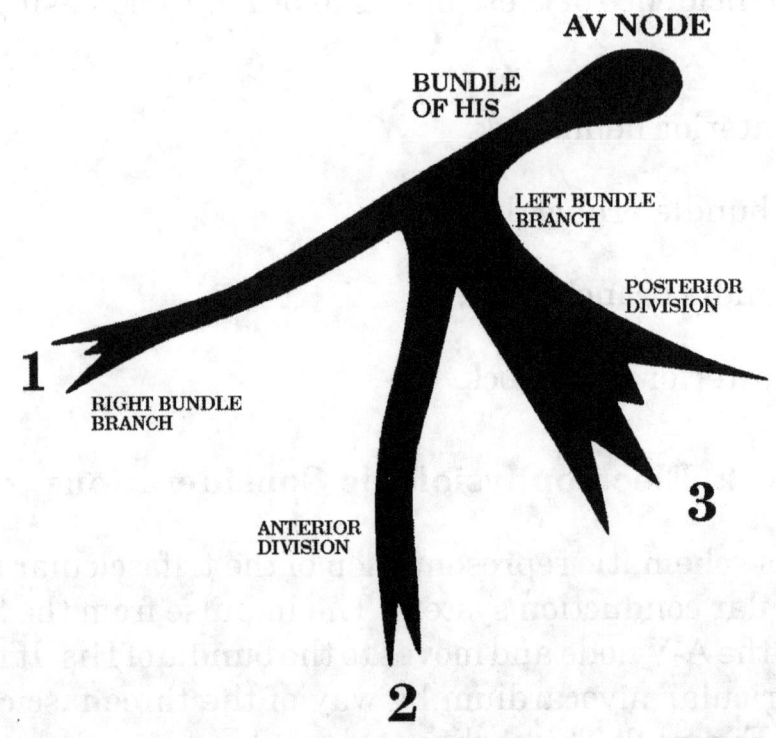

Figure 2: Trifascicular Conduction System

3. Ventricular Activation Sequence

The ventricles are depolarized in an orderly manner region by region. Ventricular depolarization begins in the interventricular septum. These forces move to the right and often superiorly, resulting in a negative deflection, or Q wave, in leads I and aVF. The activation front then moves anteriorly, laterally and posteriorly. The time that is required to complete the depolarization process averages 80 milliseconds, or 0.08 sec (Figure 3). The subsequent myocardial contraction follows the same sequence.

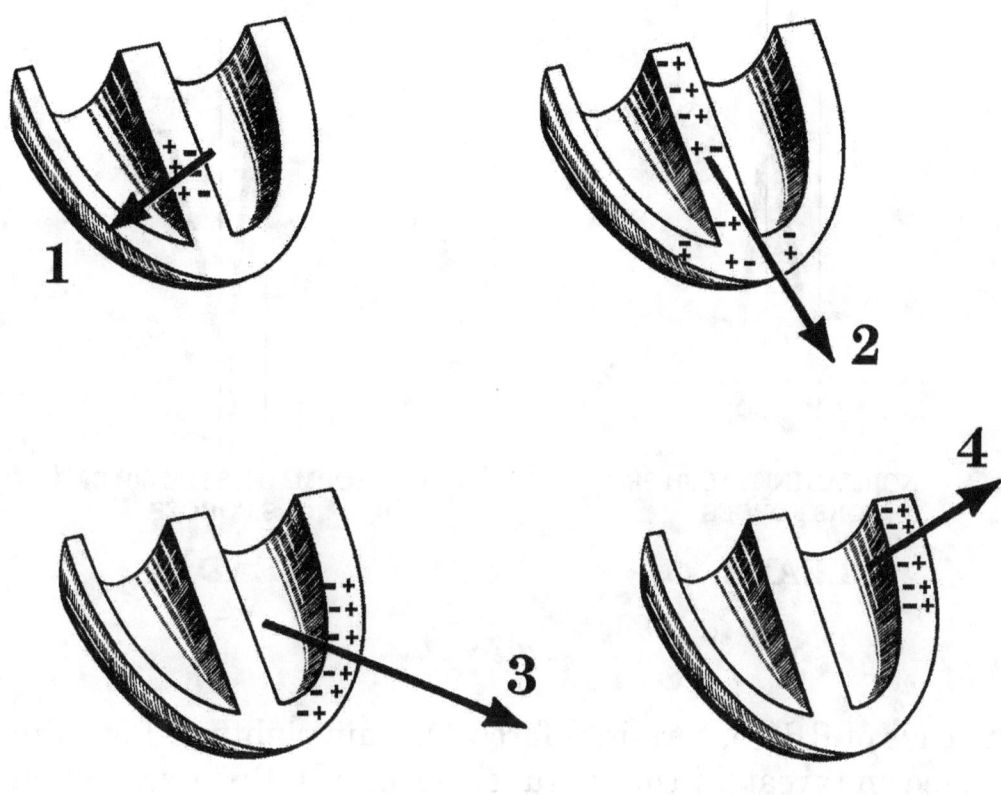

Figure 3

Normal ventricular activation results in a normal ventricular contraction. However, when the activation sequence is altered by a ventricular conduction defect, the ventricular contraction is also changed, often with adverse hemodynamic consequences.

B. Complete Right Bundle Branch Block

In complete right bundle branch block (CRBBB), the QRS interval is prolonged to 0.11 sec or more. The mean QRS vector may be directed to the left or to the right. The initial 0.04-sec of the QRS vector is *never* altered by the block. However, the terminal 0.04-sec of the QRS vector is significantly changed when CRBBB is present. Terminal QRS forces are larger in magnitude, delayed and directed rightward and anteriorly. This results in a wide negative deflection, an S wave, in lead I (Figure 5B) and a wide positive deflection, an R or R', in V_1 (Figure 6B). The terminal QRS vector in CRBBB is ***always*** directed to the right.

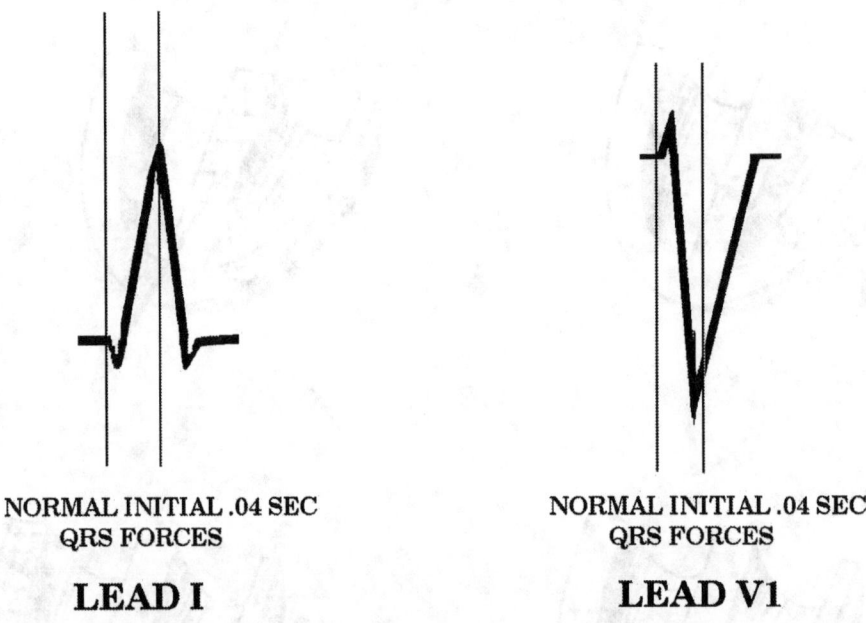

NORMAL INITIAL .04 SEC
QRS FORCES

LEAD I

NORMAL INITIAL .04 SEC
QRS FORCES

LEAD V1

Figure 4

In the case of RBBB, terminal forces remain rightward but are very prolonged and greater in magnitude. Terminal QRS forces switch from negative to positive in V_1 and are significantly prolonged (Figures 4,5 & 6).

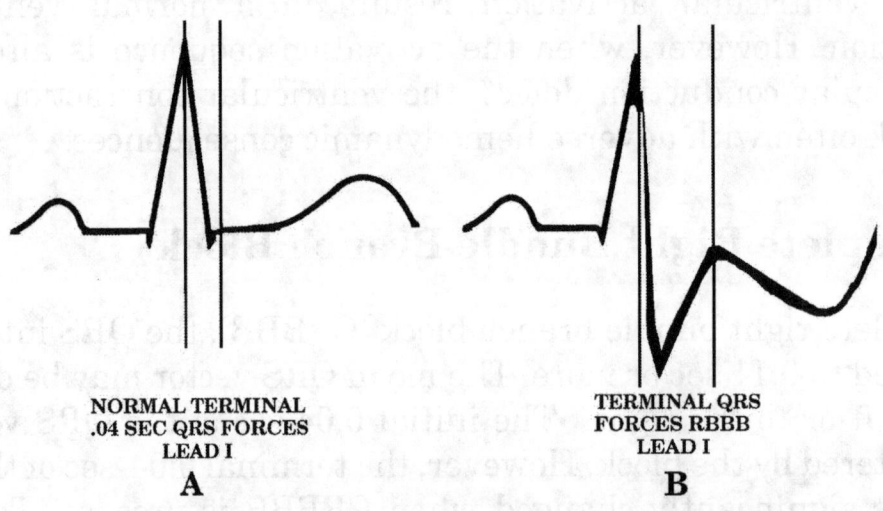

NORMAL TERMINAL
.04 SEC QRS FORCES
LEAD I
A

TERMINAL QRS
FORCES RBBB
LEAD I
B

Figure 5: Lead I

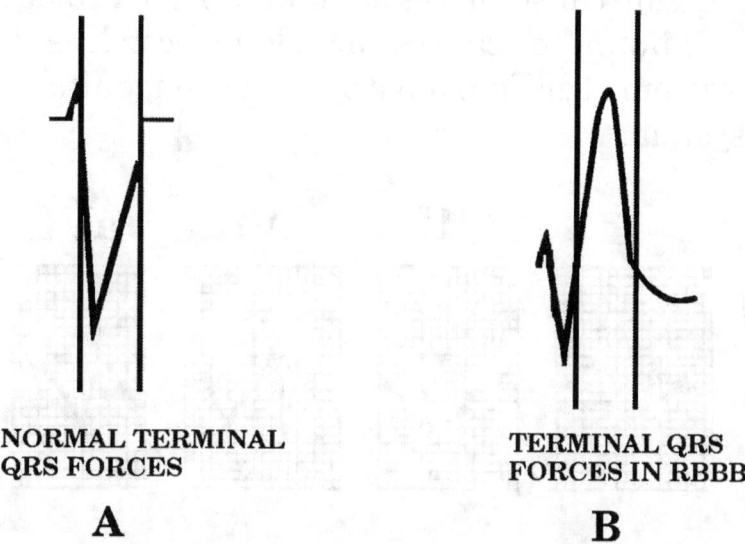

NORMAL TERMINAL QRS FORCES	TERMINAL QRS FORCES IN RBBB
A	**B**

Figure 6: Lead V₁

Right bundle branch block disrupts orderly sequential ventricular depolarization causing the left ventricle to be depolarized before the right ventricle. The T wave vector is directed opposite that of the major QRS deflection, i.e., left and posterior. The leftward and posterior direction of the T vector results in the inscription of an upright T in lead I (a leftward force) and a negative T in V_1 (a posterior force).

When terminal QRS forces are delayed and directed rightward and anteriorly and the QRS duration is greater than 0.11 sec, right bundle branch is present.

The salient features of CRBBB are:

a. QRS interval prolonged to 0.11 sec or more

b. Terminal 0.04 sec QRS forces are delayed and directed to the right and anterior

c. The T vector is directed to the left and posterior opposite the major QRS deflection of the CRBBB.

d. The mean QRS vector may be oriented to the left or right

131

The most frequent cause of right bundle branch block is coronary arteriosclerosis but other causes include myocarditis, fibrosis of the conduction system, cardiomyopathy, and congenital defect of the conduction system.

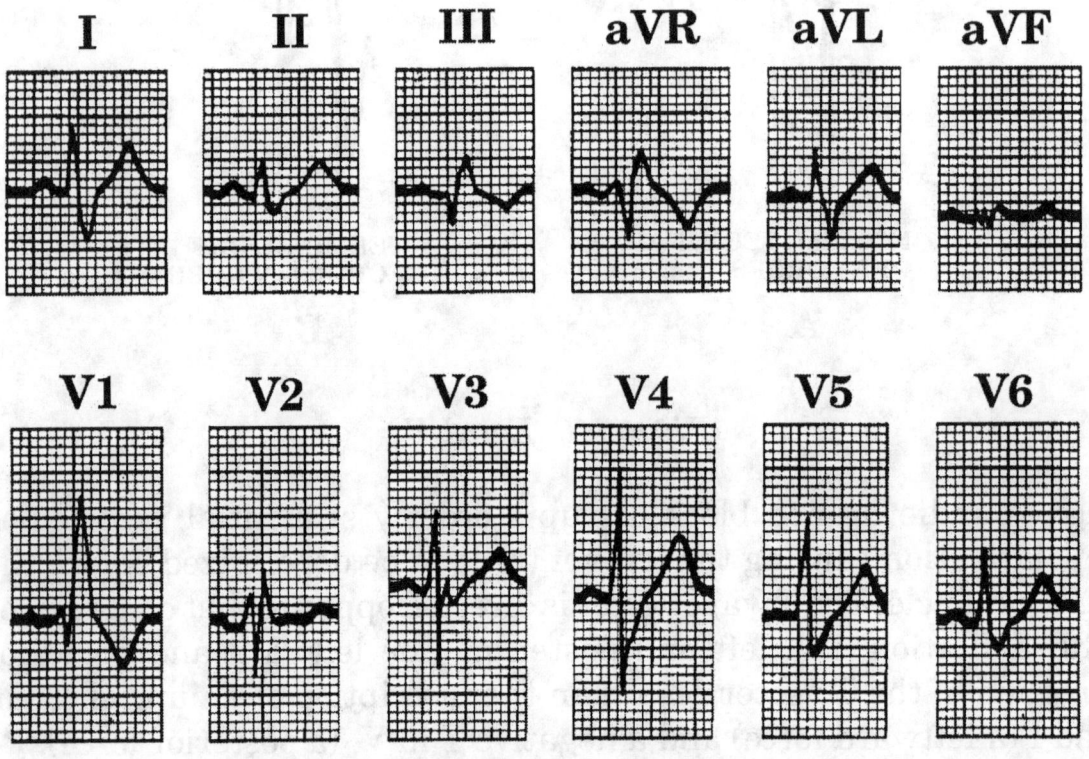

Example 1

In the tracing in Example 1, the QRS interval measures 0.12 sec indicating an intraventricular conduction defect. Terminal QRS forces are delayed and directed to the right and anterior toward the right ventricle, indicating CRBBB. This results in the inscription of a wide S in lead I (terminal rightward force) and a wide R' in V1 (terminal anterior force). The T vector is directed leftward and posteriorly as shown by the positive T wave in lead I (leftward force) and the negative T wave in V1 (posterior force).

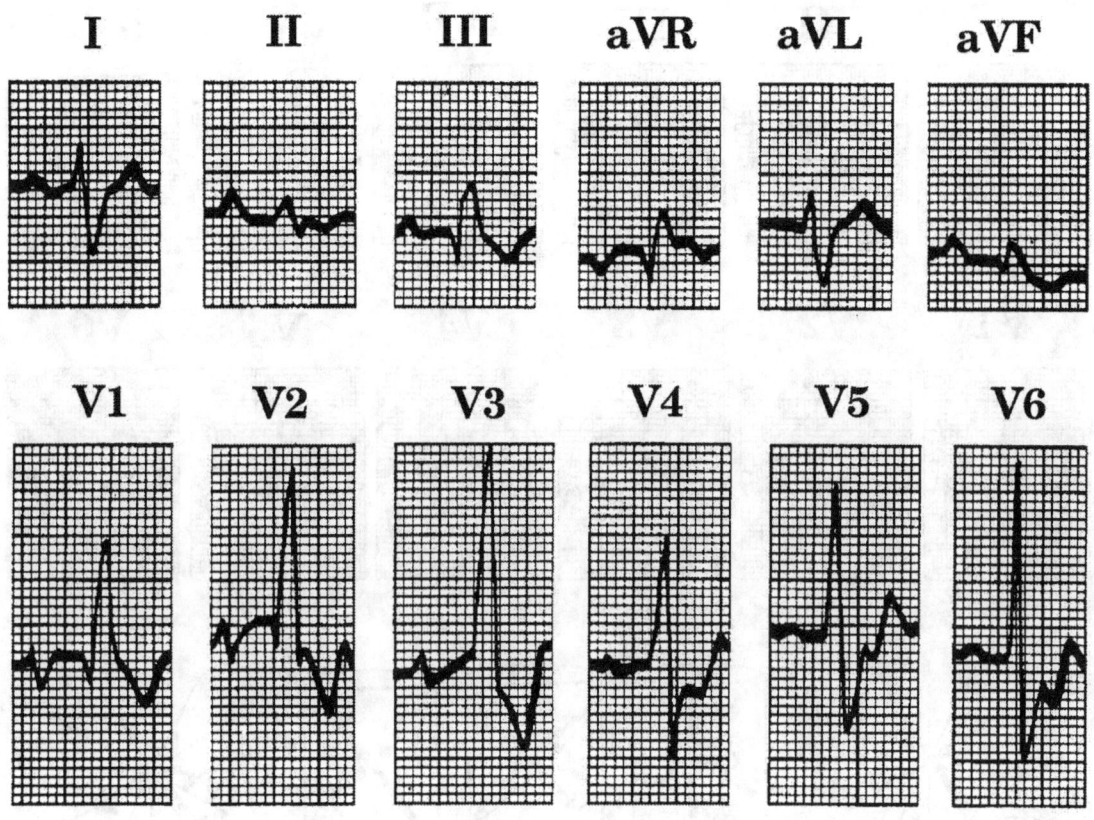

Example 2

CRBBB is present in Example 2. The QRS interval measures 0.12 sec indicating an intraventricular conduction defect. There is a terminal QRS delay oriented to the right and anterior toward the right ventricle indicating CRBBB. The T vector is directed leftward and posteriorly as shown by the upright T in lead I (leftward force) and a negative T wave in V_1 (posterior force). In contrast to the first example, there is a qR complex in V1 and not an rsR'. Nevertheless, the spatial criteria for CRBBB are satisfied since there is a terminal QRS delay oriented rightward and anterior, regardless of the qR in V_1. A Q in V_1 may be the result of an old anterior wall myocardial infarction. This will be covered in Chapter 6 on myocardial infarction.

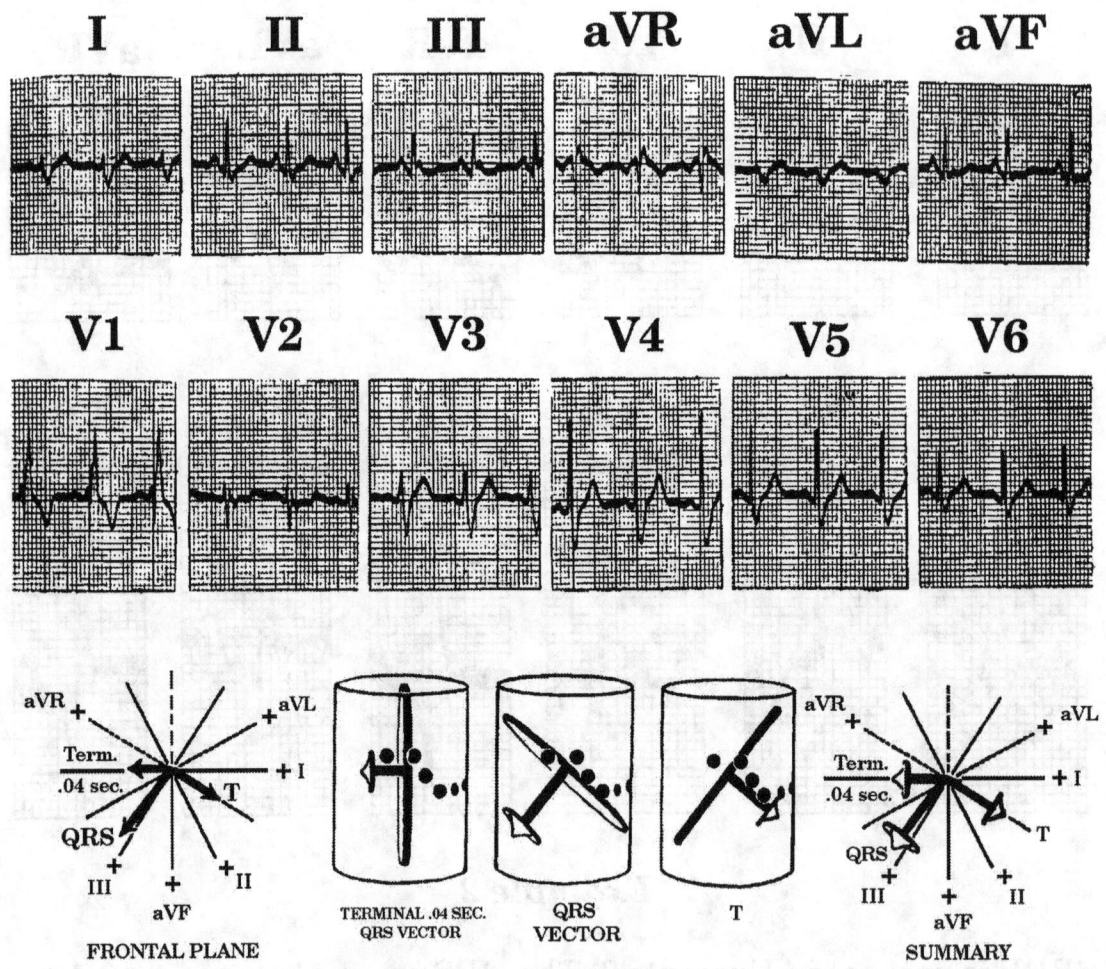

I   II   III   aVR   aVL   aVF

V1   V2   V3   V4   V5   V6

FRONTAL PLANE   TERMINAL .04 SEC. QRS VECTOR   QRS VECTOR   T   SUMMARY

Example 3

In the above EKG, all the spatial criteria for CRBBB are met. The QRS interval is 0.12 sec. The frontal plane projection shows the mean QRS vector at +130°, the mean T vector to the left at +30°, and the terminal QRS vector to the right at 180°. The terminal QRS forces are delayed and are directed to the right and anterior pointing to the right ventricle, indicating CRBBB.

C. Complete Left Bundle Branch Block

In left bundle branch block the QRS interval is prolonged to 0.12 sec or more, and the terminal 0.04-sec QRS vector points leftward and posteriorly. In contrast to CRBBB, when left bundle branch block develops, initial QRS forces are always changed in direction, pointing more leftward and more posteriorly than during normal conduction. Therefore, leads I and V_6 rarely have q waves in the presence of left bundle branch block. Also, note in Figure 7 that the mean QRS vector in both tracings is oriented essentially in the same direction, that is, to the left and inferiorly. This suggests that axis deviation has the same clinical and electrocardiographic implications during left bundle branch block that it has when the QRS interval is normal (Electrical Axis, page 109).

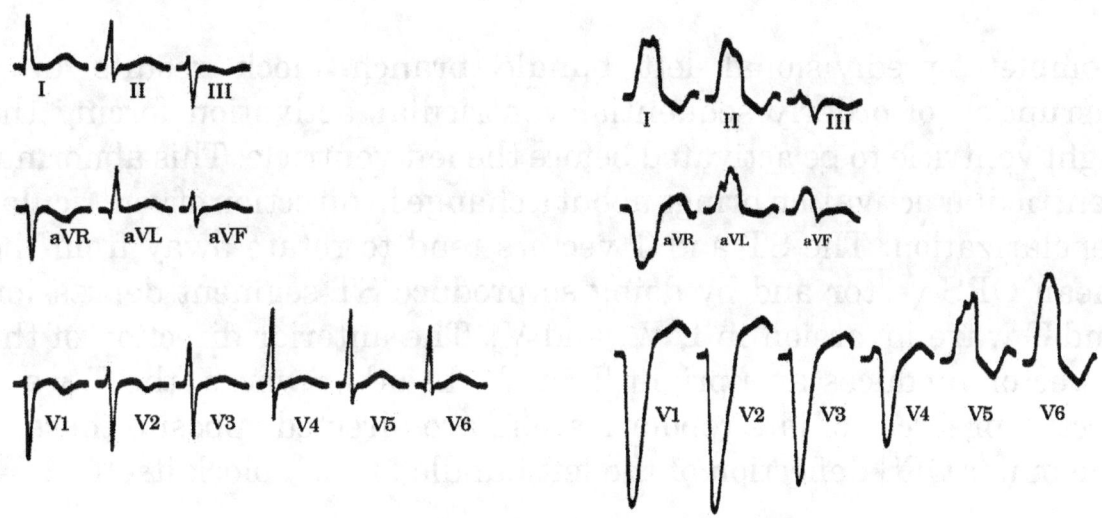

Figure 7

In Figure 7, both tracings are from the same individual. The one on the left is before the development of left bundle branch block and the tracing on the right is during block. In the tracing to the left, small Q waves are present in leads I and V_6, but in the tracing during left bundle branch block on the right, there are no Q waves. Initial QRS forces without block are oriented to the right and anteriorly, whereas, after the development of block, initial QRS forces are directed to the left and posteriorly.

In complete predivisional left bundle branch block (CLBBB), the block is located in the main left bundle. Depolarization can reach the left ventricle only by first crossing the interventricular septum from the right side to the left.

The QRS interval is prolonged to 0.12 sec or more and occasionally to 0.20 sec. The mean QRS vector is usually directed to the left but may be oriented to the right. The delay of left ventricular depolarization occurs during the terminal portion of the QRS complex and is oriented leftward, posteriorly and usually superiorly. Septal forces are often lost resulting in no Q wave in lead I or V_6. Initial QRS forces become more posteriorly oriented which results in a loss of the R wave in V_1, V_2, and sometimes in V_3. The leftward direction of the terminal QRS forces manifests itself as a wide often notched positive deflection, or R wave, in lead I. The posterior direction of the terminal QRS forces is seen as a wide negative deflection, QS or rS, in V_1.

Complete predivisional left bundle branch block results in a disruption of orderly sequential ventricular activation forcing the right ventricle to be activated before the left ventricle. This abnormal ventricular activation brings about a change in direction of ventricular repolarization. The ST and T vectors tend to rotate away from the mean QRS vector and by doing so produce ST segment depression and T wave inversion in I, V_5, and V_6. The anterior direction of the T vector produces an upright T in V_1. In other words, the T wave vector, or forces of late repolarization, are directed opposite those of the major QRS deflection of the left bundle branch block itself.

The salient features of CLBBB are:

    a. QRS interval is 0.12 sec or more in duration

    b.terminal 0.04 sec QRS forces are directed to the left and posteriorly

    c. Initial QRS vector is redirected to the left and posteriorly

The angle between the initial and mean QRS vectors rarely exceeds 45°

The most common causes of left bundle branch block are coronary atherosclerosis and fibrosis of the proximal conducting system.

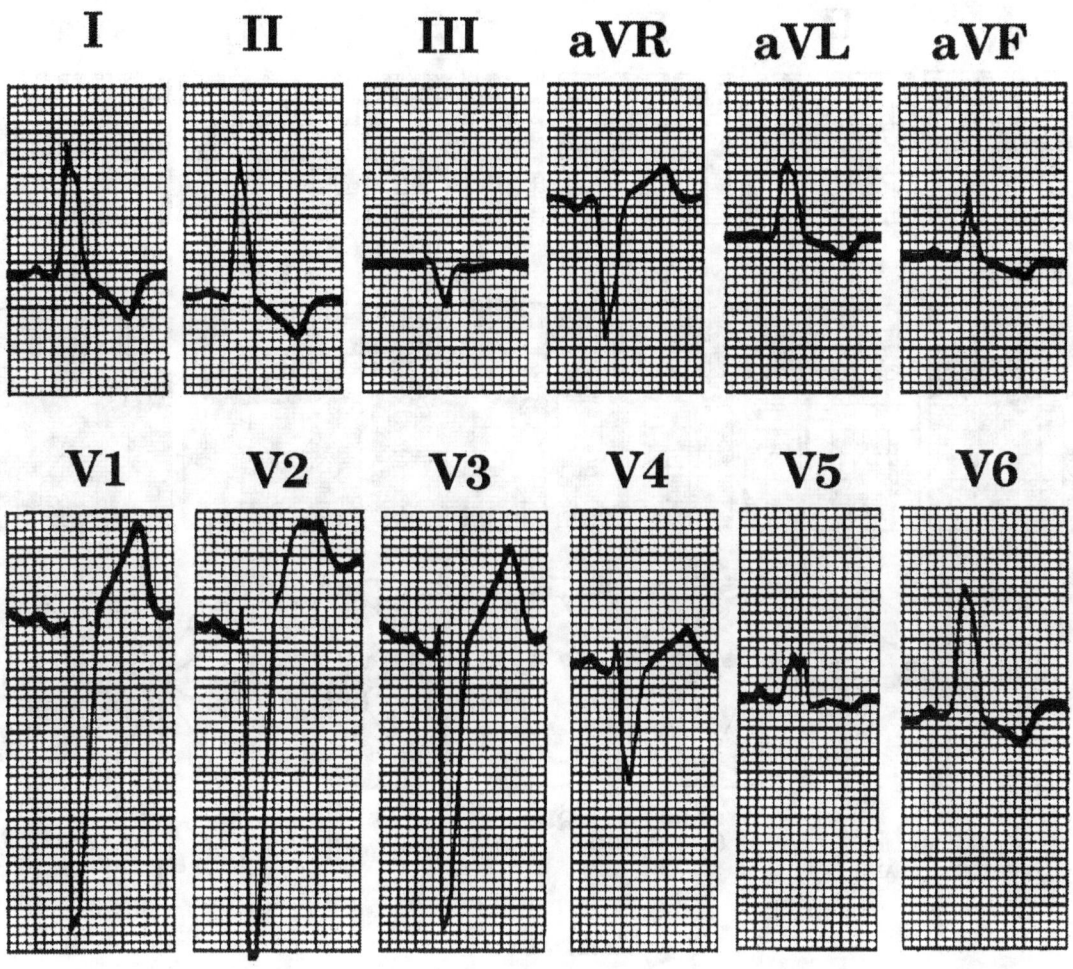

Example 4

In Example 4, the QRS interval is prolonged to 0.14 sec indicating bundle branch block. The mean QRS vector is directed to the left at +15°. The initial QRS vector is directed to the left and inferiorly at +30°. The angle between the initial and mean QRS vectors is 20°. Terminal QRS forces are delayed and oriented to the left and posteriorly pointing to the left ventricle, indicating CLBBB. This is shown by the notched R wave in lead I (leftward force) and a large wide negative deflection in V1 (posterior force). The ST–T vectors are directed to the right and anteriorly. These forces are responsible for the inscription of the depressed ST segment and negative T in lead I (rightward forces) and elevated ST and upright T in V_1 (anterior forces).

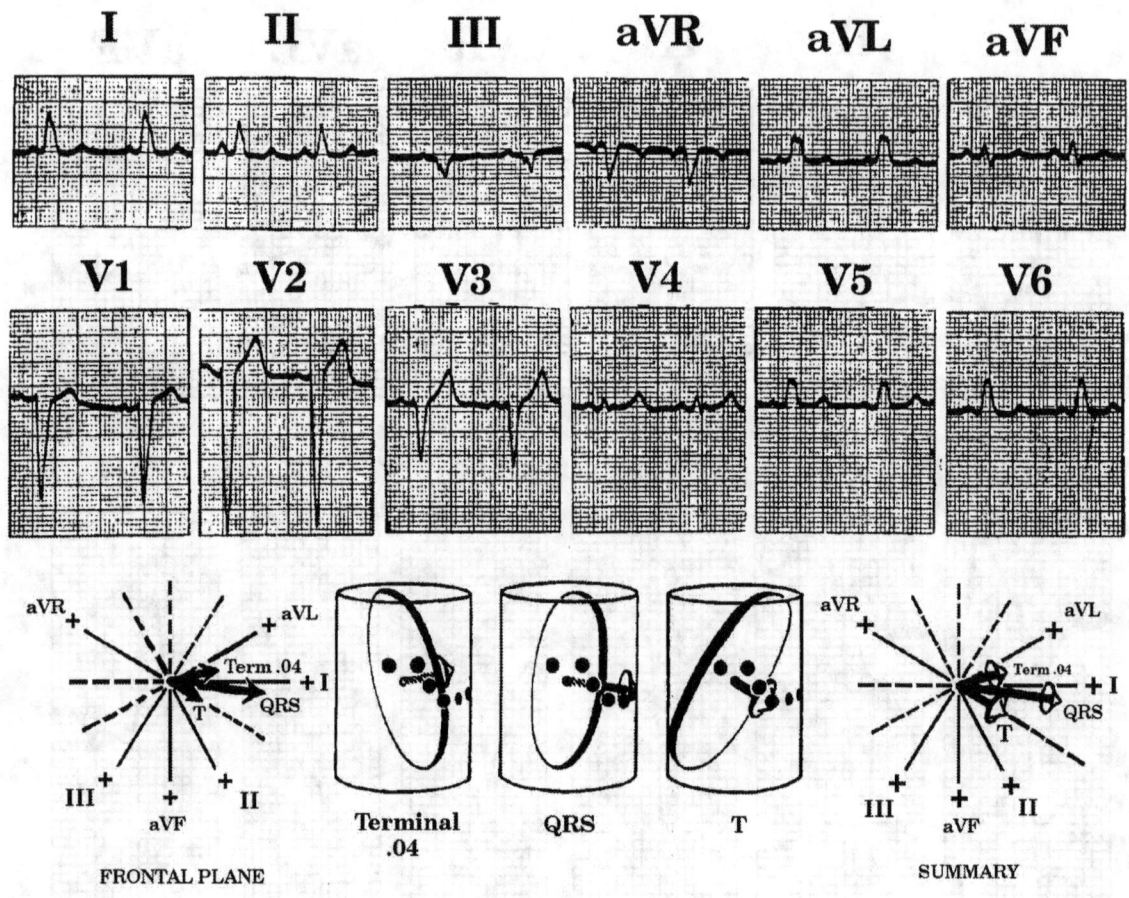

Example 5

In Example 5, the spatial criteria for CLBBB have been met. The QRS interval is 0.13 sec indicating bundle branch block. The frontal plane projection shows the mean QRS vector at + 5°, the mean T vector at +30°, and the terminal 0.04 sec vector at –15°. The terminal QRS forces are delayed and are directed to the left and posterior pointing toward the left ventricle and, therefore, indicate LBBB.

D. The Hemiblocks

1. Historical Considerations

In 1934, Wilson reported three cases of right bundle branch block with superior axis deviation (see Chapter 3, page 110, "Axis Deviation"). The QRS complexes in leads II and III were of the rS configuration. Wilson decided to compare his three cases with a case reported by

Mahaim in which there was a similar electrocardiogram. The autopsy of Mahaim's case revealed an anteroseptal myocardial infarction and injury to both the right bundle branch and the anterior division of the left bundle branch. Wilson felt that the cause of the rS complexes in leads II and III was an interruption of conduction in the anterior division of the left bundle branch. This was the first time anyone considered the possibility of anterior division block as an explanation for the superior axis deviation in the human EKG.

In 1950, Rosenbaum et al. saw a patient who had sustained an anterior myocardial infarction. Here the EKG showed right bundle branch block and a mean frontal QRS vector of −75°. Several weeks later the same patient had an EKG that showed right bundle branch block and a mean frontal QRS vector of +110°. Serial tracings of this patient showed right bundle branch block and a mean frontal QRS vector of −75° that alternated with a mean frontal QRS vector of +110°. Rosenbaum concluded that in this patient the left ventricle was activated before the right ventricle regardless of the location of the mean frontal QRS vector since right bundle branch block was present in both instances.

To verify his suspicions, Rosenbaum asked his colleagues to examine the anatomy of the ventricular conduction system. With careful dissection, they found that the main left bundle branch divided early into an anterior and posterior division. Rosenbaum correctly reasoned that an alternating block of the anterior and posterior divisions of the left bundle branch caused the switch in the mean frontal QRS vector seen in his patient.

The mechanism of hemiblock can be explained further in terms of the anatomy and electrophysiology of the left ventricular conducting system. Shortly after its origin, the left bundle branch divides into two groups of fibers. One group, called the anterior division, radiates anteriorly and superiorly to reach the anterolateral wall of the left ventricle. The other group, called the posterior division, is distributed to the inferoposterior wall of the left ventricle. **Recall that any block requires excitable tissue beyond it to be expressed electrocardiographically.**

2. Left Anterior Superior Hemiblock

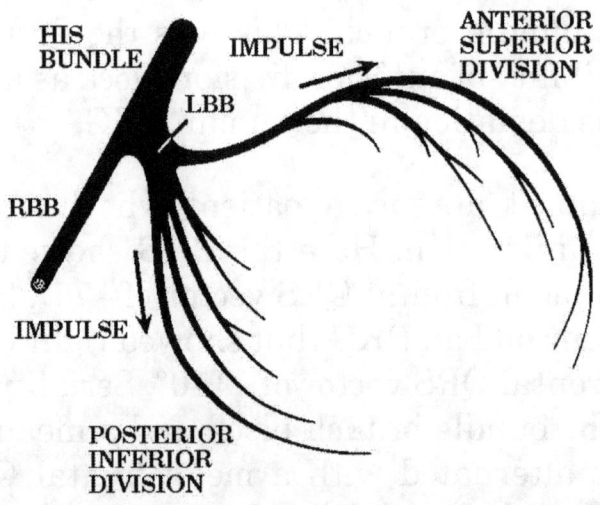

HIS BUNDLE

IMPULSE

ANTERIOR SUPERIOR DIVISION

LBB

RBB

IMPULSE

POSTERIOR INFERIOR DIVISION

NORMAL CONDUCTION

Figure 5

Excitation normally spreads simultaneously through both divisions (Figure 5). A delay or failure of impulse conduction through the anterior division of the left bundle branch will cause asynchronous activation of the left ventricle. Left ventricular wall motion will be abnormal. Simultaneous interruption of impulse conduction through both divisions would have the same effect as block of the main left bundle branch. The electrocardiographic abnormalities would be indistinguishable from complete left bundle branch block.

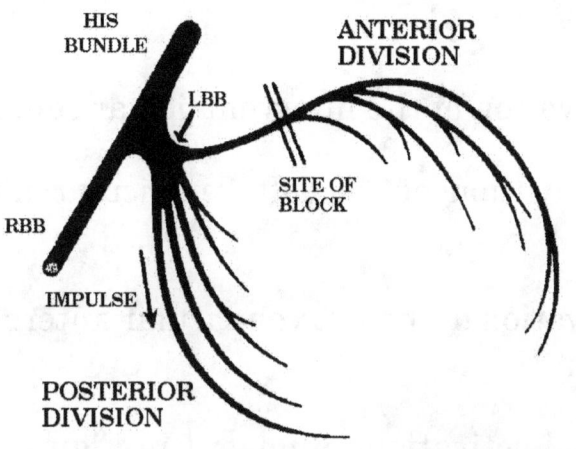

LEFT ANTERIOR HEMIBLOCK

Figure 6

When the anterior division becomes blocked, impulses must travel through the fibers of the posterior division and spread superiorly. The terminal QRS vector will be shifted to the left and superiorly and left axis deviation will occur, **with or without** prolongation of the QRS interval (Figure 6).

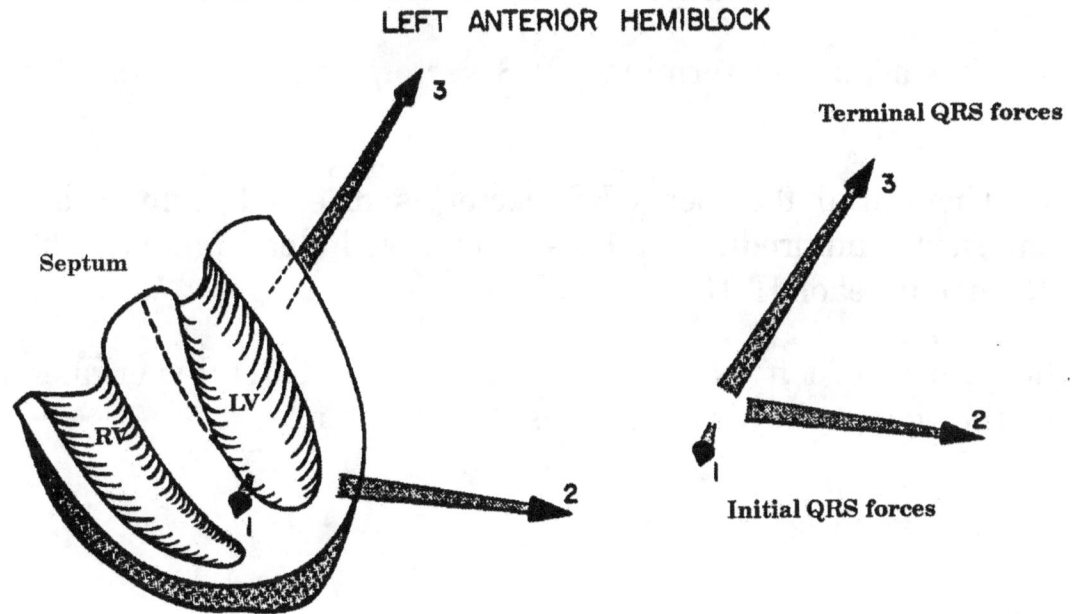

Figure 7

The ventricular activation process in left anterior hemiblock may be divided three parts:

a. Phase 1, activation of the interventricular septum

b. Phase 2, activation of the left ventricular inferior wall and apex

c. Phase 3, activation of the left ventricular anterolateral wall and posterior region,

The vector of septal activation, number 1 in Figure 7, is directed to the right and inferiorly. These septal forces will cause an inscription of a small Q wave in lead I (rightward force) and a small R wave in leads II, III, and aVF (inferior force).

The vector of activation of the inferior and posterobasal region of the left ventricle, number 3 in Figure 7, is directed leftward and superiorly. These forces will produce a mean frontal QRS vector usually between −45° and −75°.

3. Criteria for Left Anterior Hemiblock

a. The mean and terminal QRS vectors are shifted to −45° or more

b. The initial 0.02-sec QRS vector is directed rightward and inferiorly and produces a 0.02-sec Q wave in lead I and a 0.02-sec R wave in leads II, III, and aVF.

Other causes of left axis deviation, such as, COPD, hyperkalemia, and other metabolic disorders must be ruled out.

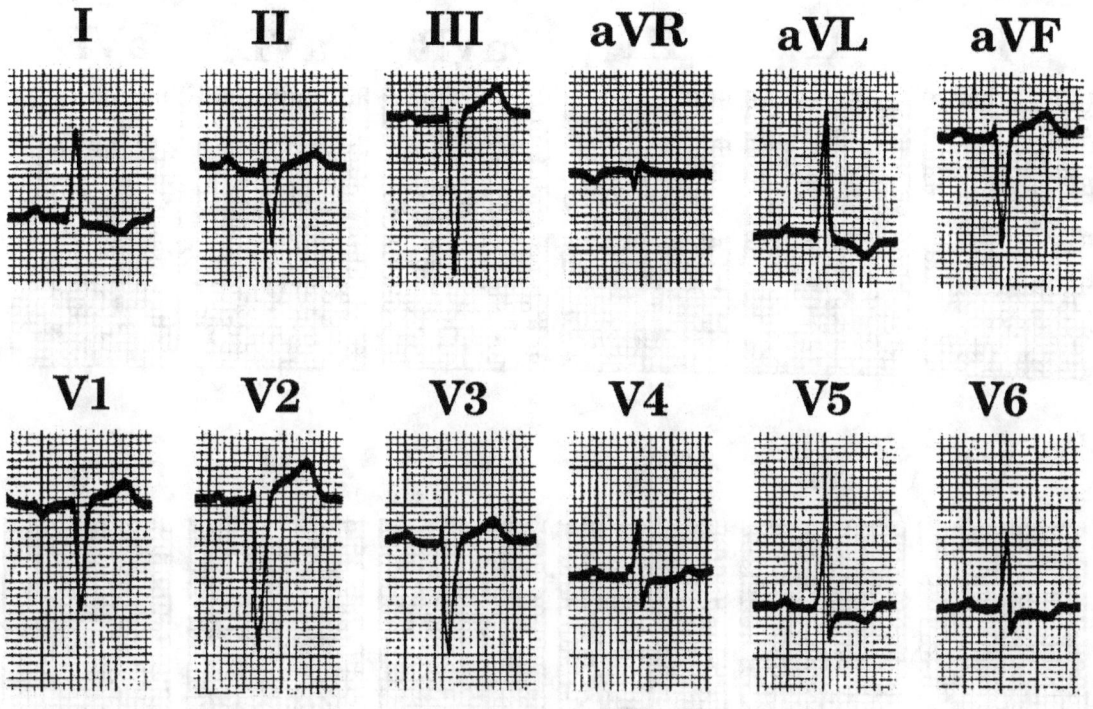

Example 6

In Example 6, a predominately positive deflection in lead I indicates QRS forces are moving to the left. Lead aVF has a small positive deflection followed by a much larger negative deflection consistent with QRS forces directed superiorly. The upright QRS deflection in lead I combined with the negative QRS deflection in aVF places the mean frontal QRS vector in the left superior quadrant.

Lead aVR displays a nearly equiphasic QRS complex which indicates that the mean QRS vector is close to being perpendicular to the axis of aVR in the left superior quadrant. This locates the mean frontal QRS vector at –60°.

Initial QRS forces are oriented to the right and inferiorly as indicated by the small Q wave in lead I and the small R wave in aVF. The criteria for left anterior hemiblock have been met.

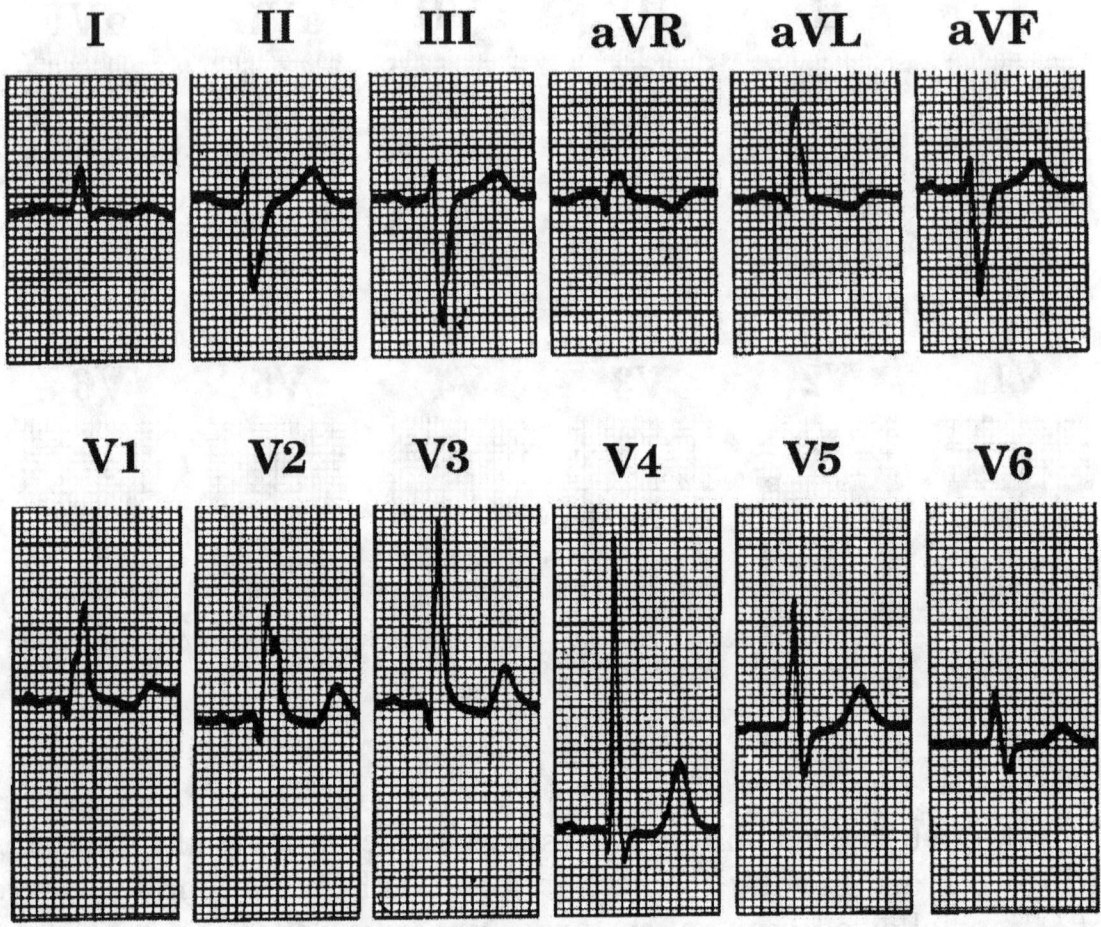

Example 7

In Example 7, initial QRS forces are oriented to the right and inferiorly as shown by the small Q wave in lead I and small R wave in aVF. There is an R wave in lead I meaning QRS forces are moving to the left. Lead aVF has a large negative deflection consistent with QRS forces moving superiorly. The upright QRS deflection in lead I combined with the negative QRS deflection in aVF places the mean frontal QRS vector in the left superior quadrant.

There is no lead with an equiphasic complex. The QRS complex in Lead aVR is predominately positive and is largest and negative in lead III. This indicates that the mean frontal QRS vector is greater than –60° but not quite –90°. This would place the mean frontal QRS vector somewhere between –60° (perpendicular to the axis of lead III) and –90° (perpendicular to the axis of aVF), or about –75°. The QRS

144

interval measures 0.12 sec and terminal QRS forces are oriented to the right and anteriorly because of a broad S wave in lead I and a large notched R wave in V1. These findings are consistent with complete right bundle branch block. Furthermore, the criteria for left anterior hemiblock have been met. The final diagnosis is complete right bundle branch block and left anterior hemiblock.

4. Left Posterior Hemiblock

When the posterior division is blocked, excitation must move through the fibers of the anterior division and spread inferiorly (Figure 8). The terminal QRS vector will be directed inferiorly and to the right resulting in right axis deviation (see Chapter 3, page 114) with or without prolongation of the QRS interval.

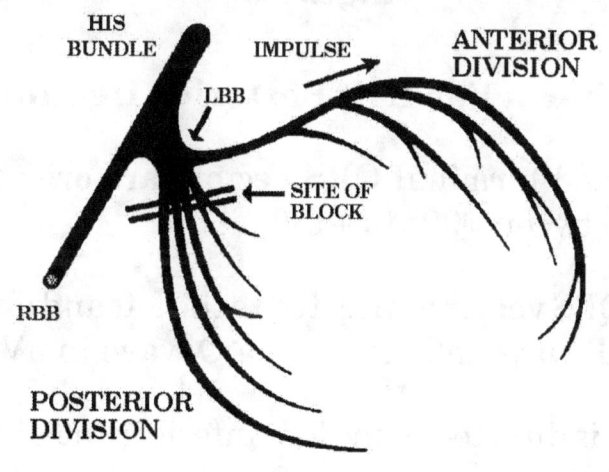

LEFT POSTERIOR HEMIBLOCK

Figure 8

Left posterior hemiblock is less common than left anterior hemiblock. Rosenbaum attributes the lower incidence to the following characteristics of the posterior division:

a. The fibers of the posterior division are short and thick

b. It has a dual blood supply

c. It lies in the less turbulent inflow tract of the left ventricle

145

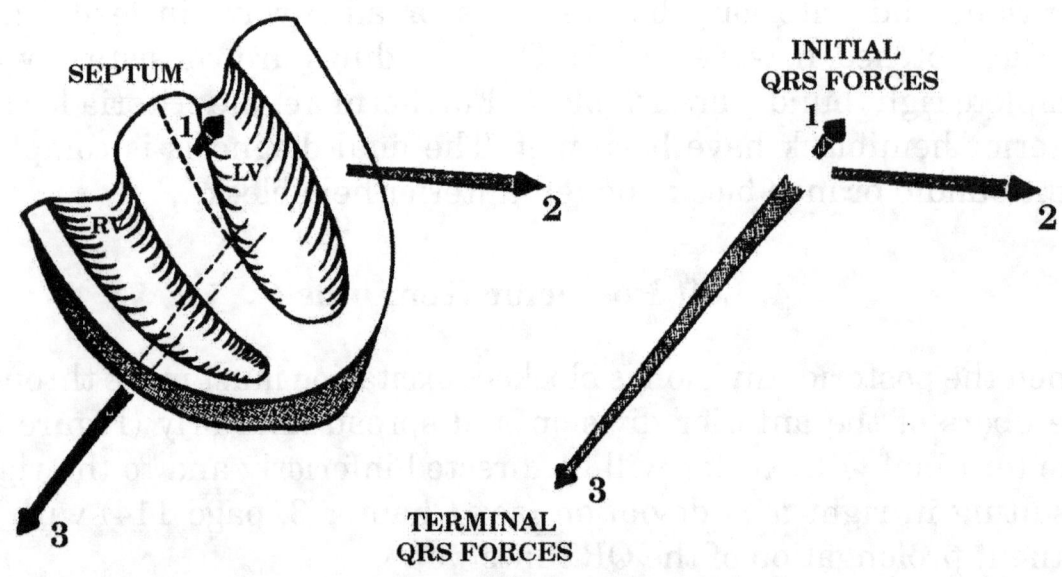

Figure 9

5. Criteria For Left Posterior Hemiblock

a. The mean and terminal QRS vectors are oriented to the right and inferior between +90° to +130°

b. The initial QRS vector is directed to the left and always superiorly producing an R wave in lead I and a Q wave in aVF

c. The T vector is directed to the left, inferior, and anterior, resulting in positive T waves in leads I, aVF, and V_1

The septal activation vector, shown as number 1 in Figure 9, is directed to the left and superiorly. These septal forces will cause an inscription of a small Q wave in lead I (rightward force) and a small R wave in leads II, III, and aVF (inferior force).

The terminal QRS vector, shown as number 3 in Figure 9, is directed to the right and inferiorly. These forces will produce mean and terminal QRS vectors between +90° and +130°.

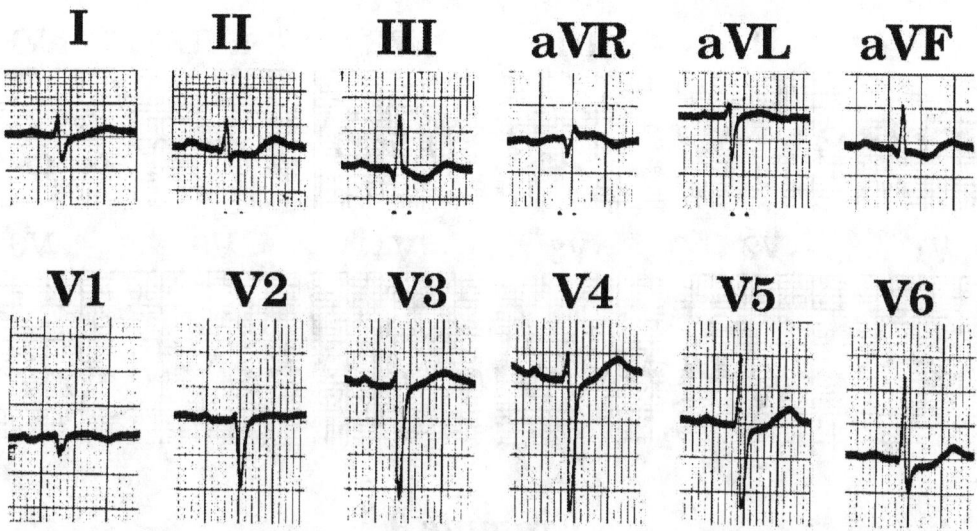

I   II   III   aVR   aVL   aVF

V1   V2   V3   V4   V5   V6

Example 8

In Example 8, a negative QRS deflection in lead I indicates forces are moving to the right. Lead aVF has a positive QRS deflection consistent with QRS forces directed inferiorly. The negative QRS deflection in lead I combined with the positive QRS deflection in aVF places the mean frontal QRS vector in the right inferior quadrant.

Lead aVR has an equiphasic QRS complex which indicates that the mean QRS vector is perpendicular to the axis of aVR in the right inferior quadrant. This places the mean frontal QRS vector at +120°.

The T vector is oriented to the left, inferiorly, and anteriorly because the T waves in leads I, aVF, and V_1 are positive. Initial QRS forces are oriented to the left and superiorly as indicated by the small R wave in lead I and the small Q wave in aVF. The criteria for left posterior hemiblock have been met.

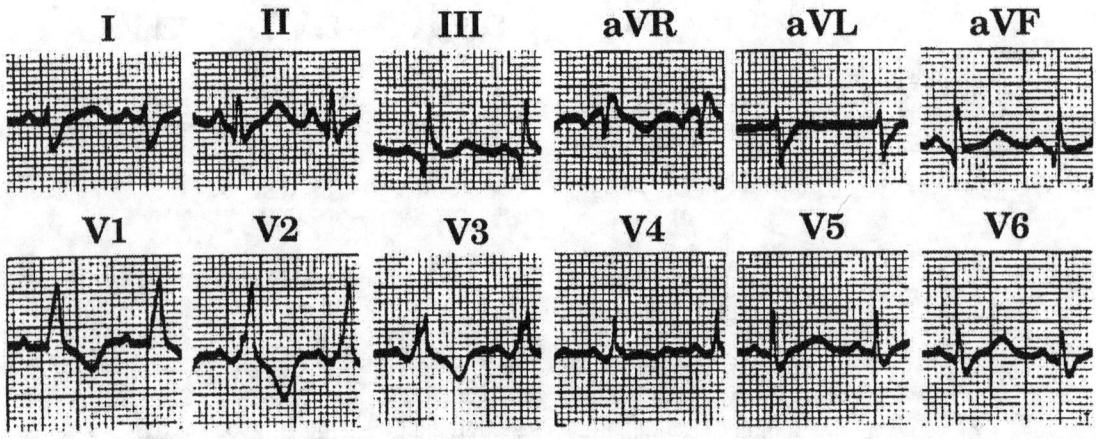

Example 9

In Example 9, initial QRS forces are oriented to the left and superiorly as indicated by the small R wave in lead I and the Q wave in aVF. There is a wider than normal S wave in lead I meaning QRS forces are moving to the right. Lead aVF has a negative deflection followed by a larger positive deflection consistent with QRS forces moving inferiorly. The negative QRS deflection in lead I combined with the positive QRS deflection in aVF places the mean frontal QRS vector in the right inferior quadrant. There is no lead with an equiphasic complex but the QRS complex in Lead aVR comes close. This means that the mean frontal QRS vector is located in the right inferior quadrant at about +120°. The QRS interval measures 0.12 sec and terminal QRS forces are prolonged and oriented to the right and anteriorly because of the wide S wave in lead I and a large wide R wave in V_1. These findings are consistent with complete right bundle branch block. Furthermore, since the criteria for left posterior hemiblock have been met, the final diagnosis is complete right bundle branch block and left posterior hemiblock (bifascicular block).

CHAPTER SIX:
ISCHEMIA, INJURY & INFARCTION

A. General Considerations

To help identify and understand electrocardiographic abnormalities caused by coronary artery occlusion, laboratory experiments were conducted on dogs. A coronary artery was tied off and electrodes were placed on the area of the heart muscle supplied by the occluded vessel. Within several minutes T waves became inverted. Coronary blood flow was immediately reestablished by removing the ligature and the inverted T waves slowly returned to normal. It was concluded that T wave inversion represented myocardial ischemia and was reversible (Figures 1,2).

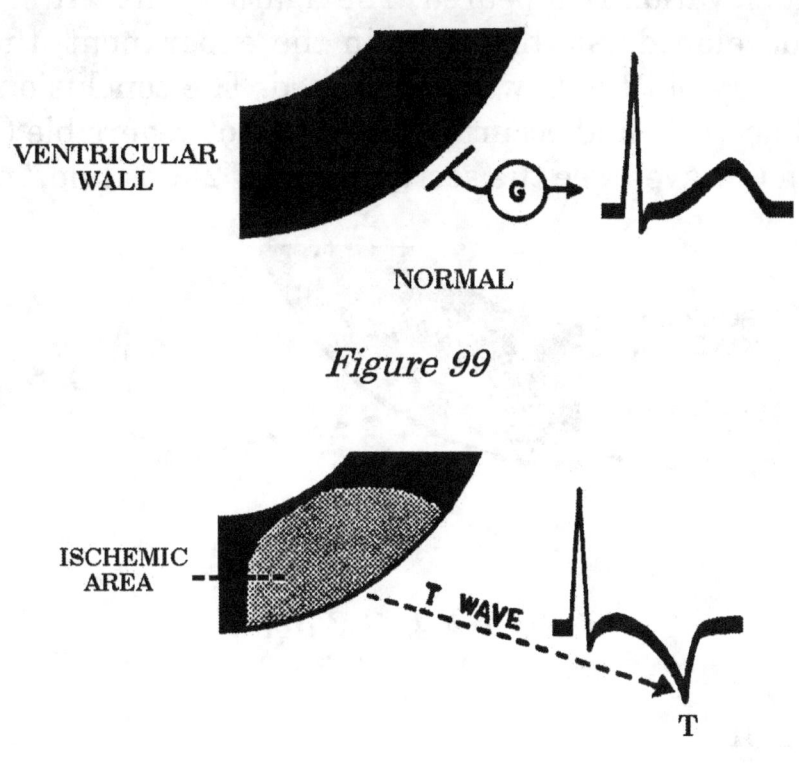

NORMAL

Figure 99

Figure 1

The coronary artery was tied off again and the ligature left in place beyond the time of the appearance of inverted T waves. Within a few minutes, the ST segments became elevated, and the tie was removed. The tracing gradually returned to normal as the ST segment returned to the baseline and the inverted T waves became upright. It was concluded that ST segment elevation represented myocardial injury, a stage beyond ischemia, and was reversible (Figure 3).

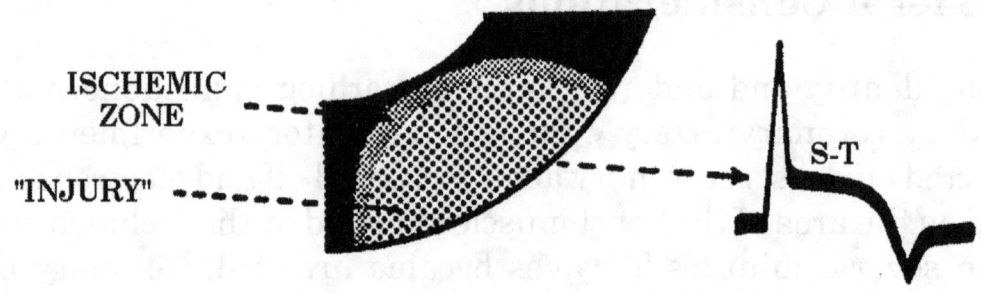

ISCHEMIC ZONE

"INJURY"

S-T

Figure 3: ST Elevation

The coronary artery was again tied off and T wave inversion and ST segment elevation reappeared. The ligature was left in place and a Q wave developed. At this point in the experiment, the ligature was taken away but the Q wave remained. The conclusion was that myocardial necrosis had occurred and was not reversible (Figure 4). For more on Q waves, see the section on the "2-3-4 Rule," page 174.

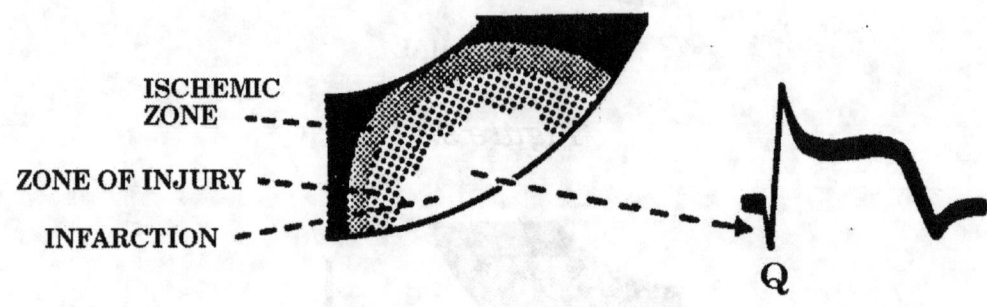

ISCHEMIC ZONE

ZONE OF INJURY

INFARCTION

Q

Figure 4: Q Wave

B. Ischemia

1. T Wave Abnormalities

The effect of myocardial ischemia on the T wave vector depends on whether the ischemic process within the ventricular wall involves the epicardial or endocardial region. When there is no myocardial ischemia, the T wave vector tends to parallel the QRS vector. The QRS-T angle is normal in this case (Figure 5).

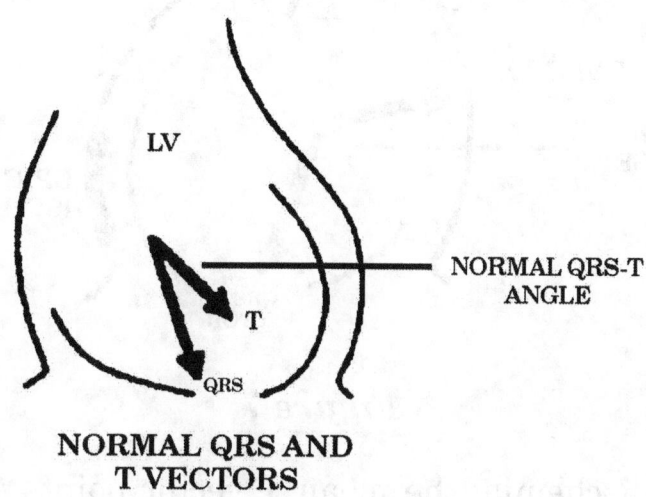

**NORMAL QRS AND
T VECTORS**

Figure 5

In epicardial ischemia, the mean T vector points *away* from the site of ischemia (Figure 6).

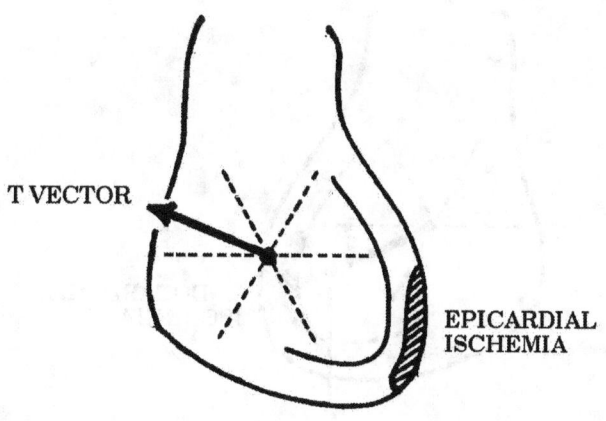

Figure 6

In myocardial infarction, ischemia is localized to the epicardial region of the involved region of the left ventricular wall. Ischemic myocardium cannot repolarize normally and the mean T vector is directed away from the site of epicardial ischemia. This shift in the T wave vector creates a wider than normal QRS-T angle (Figure 7).

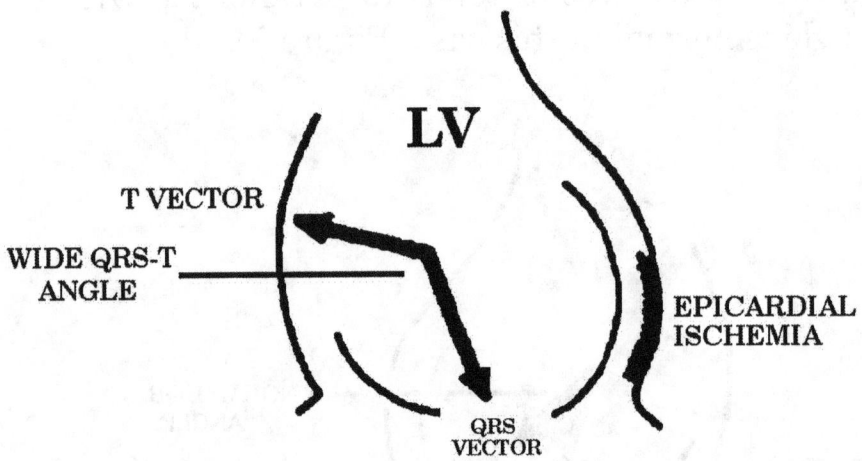

Figure 7

In endocardial ischemia, the mean T vector points *toward* the site of ischemia (Figure 8). The size of the QRS-T angle in endocardial ischemia is determined by the locations of the mean QRS vector and ventricular ischemia In most cases of endocardial ischemia, the QRS-T angle is normal (Figure 9).

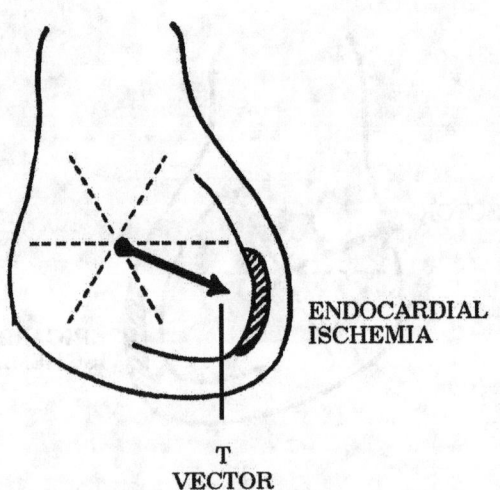

Figure 8

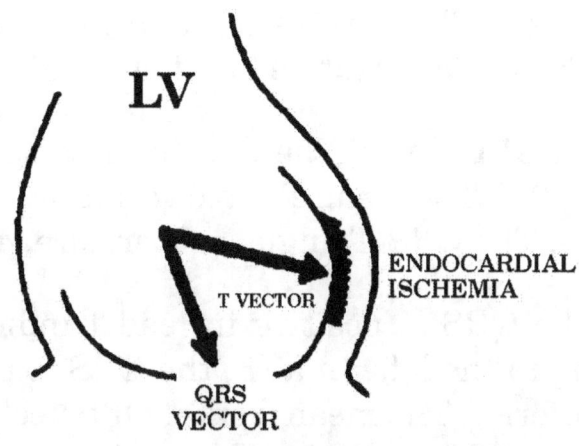

LV

T VECTOR

ENDOCARDIAL
ISCHEMIA

QRS
VECTOR

Figure 9

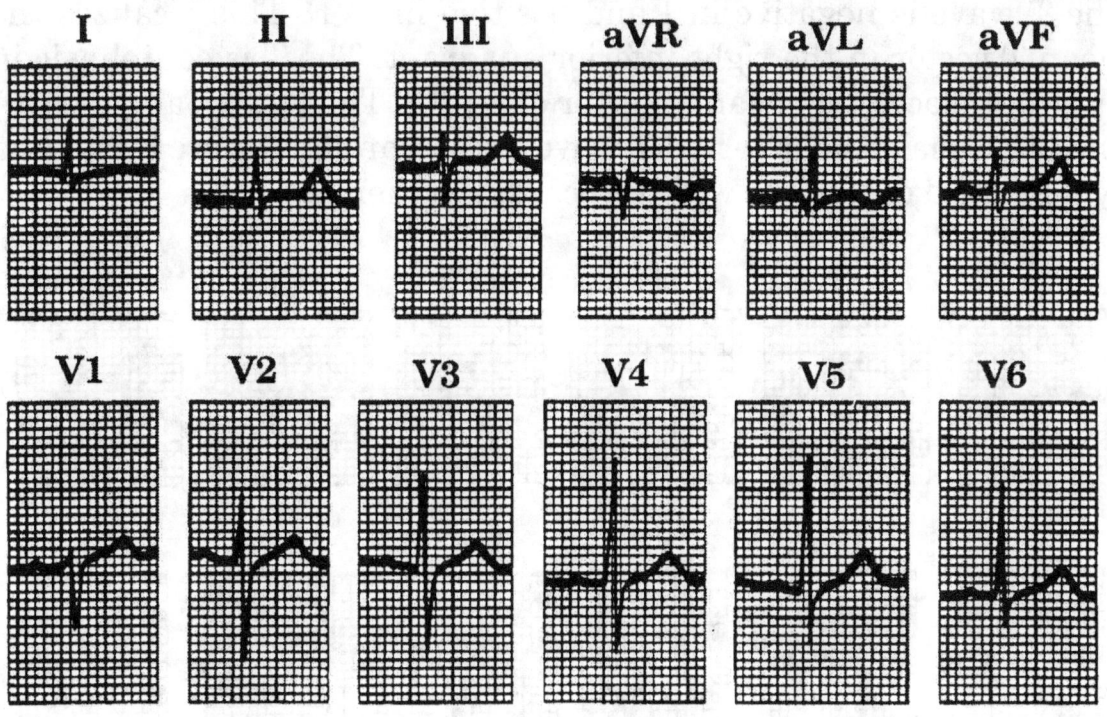

Example 1: Before Onset of Ischemia

In Example 1, the positive QRS deflection in lead I indicates QRS forces are moving to the left. In aVF, the QRS deflection is more positive than negative indicating QRS forces are inferior. The mean frontal QRS vector is localized in the left inferior quadrant. The QRS in III is equiphasic making the mean QRS vector perpendicular to the axis of III, or +30°.

The T wave is positive in I and aVF placing the mean T vector in the left inferior quadrant. The T is negative in aVL placing the mean T vector beyond +60° in the negative half field of aVL.

The T is positive in aVF placing the T vector in the positive half field of aVF at less than +90°. The mean T wave vector must lie between +60° and +90°, or +75°. The QRS-T angle is, therefore, 45°, or normal.

In Example 2, the QRS is positive in lead I indicating those QRS forces are moving to the left. In aVF, the QRS is positive indicating QRS forces are inferior. The mean frontal QRS vector has to be in the left inferior quadrant. The QRS in III is equiphasic making the mean QRS vector perpendicular to the axis of III, or +30°.

The T wave is negative in I and positive in aVF. This localizes the mean T vector in the right inferior quadrant. The T is equiphasic in aVR and places the mean T vector at +120°. The QRS-T angle is 90° and abnormal. The inverted T waves in V_1 through V_6 are broad and symmetrical consistent with myocardial ischemia.

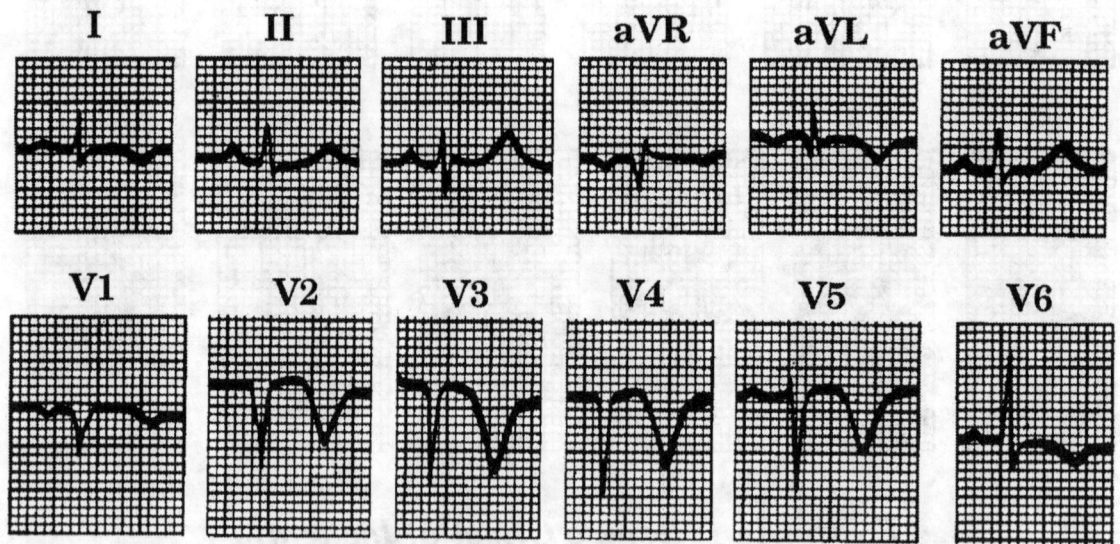

Example 2: After Onset of Ischemia

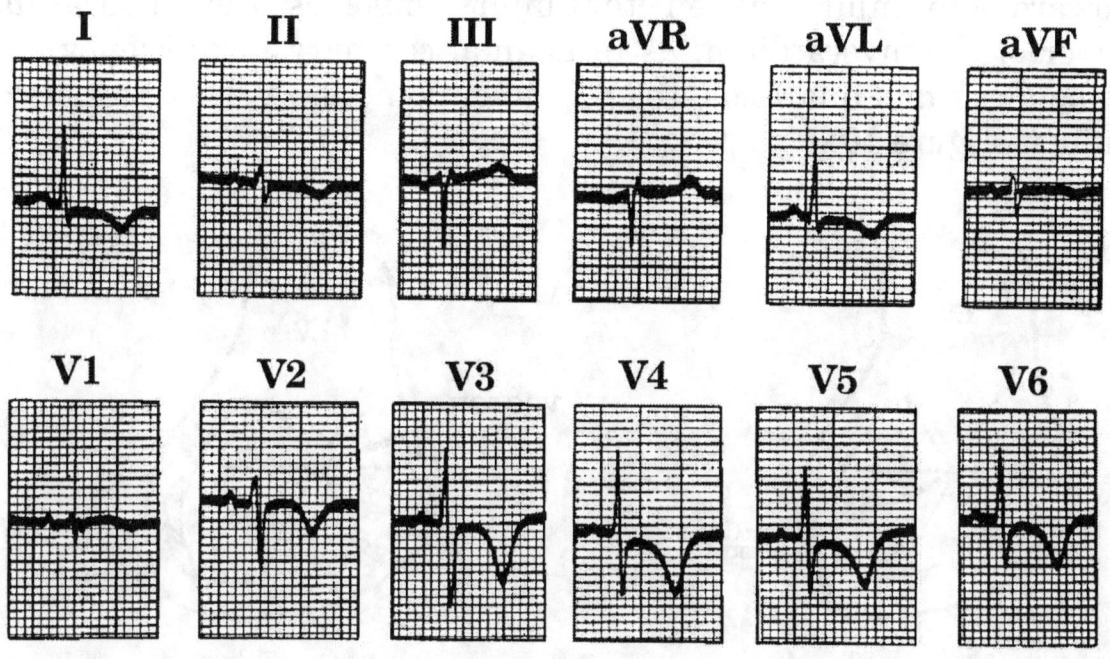

Example 3

In Example 3, the QRS is positive in I and equiphasic in aVF. This places the mean QRS vector to the left and perpendicular to. lead aVF, at 0°. The T wave is negative in I and equiphasic in aVF placing the mean T vector to the right and perpendicular to aVF, or +180°. The QRS-T angle is 180° indicating the T waves in this tracing are abnormal. The T waves in V_2 through V_6 are inverted, broad and symmetrical, consistent with myocardial ischemia.

C. Injury

1. ST Segment Abnormalities

There is a type of myocardial injury which is seen on the EKG solely as displacement of the ST segment. This is thought to be injury intermediate in severity between that which produces T vector abnormalities and frank myocardial necrosis resulting in QRS vector abnormalities.

When the injury is located in the epicardial layer of the myocardium, as in myocardial infarction and pericarditis, the ST vector points toward the injury. When the injury involves the endocardial layer of the myocardium, as in angina, coronary insufficiency and subendocardial infarction, the ST vector points away from the site of injury (Figure 10).

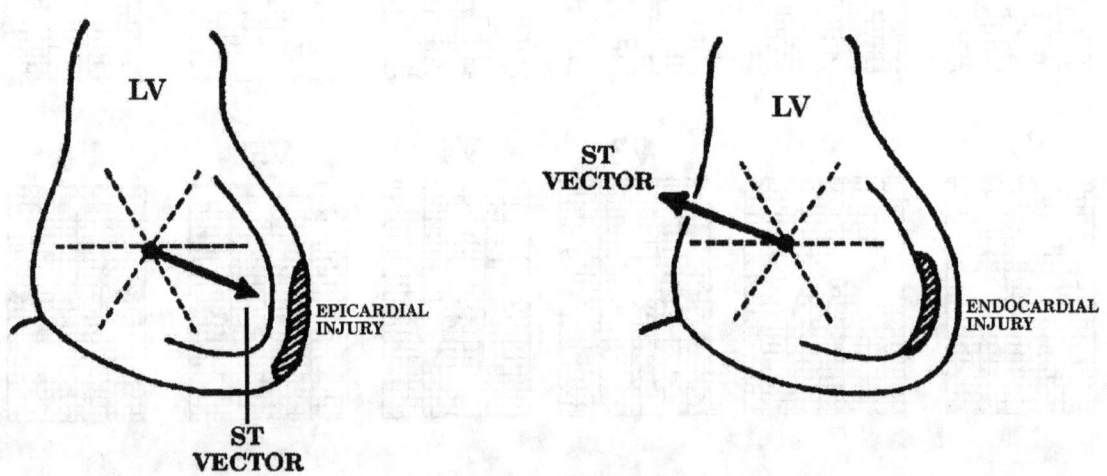

Figure 10

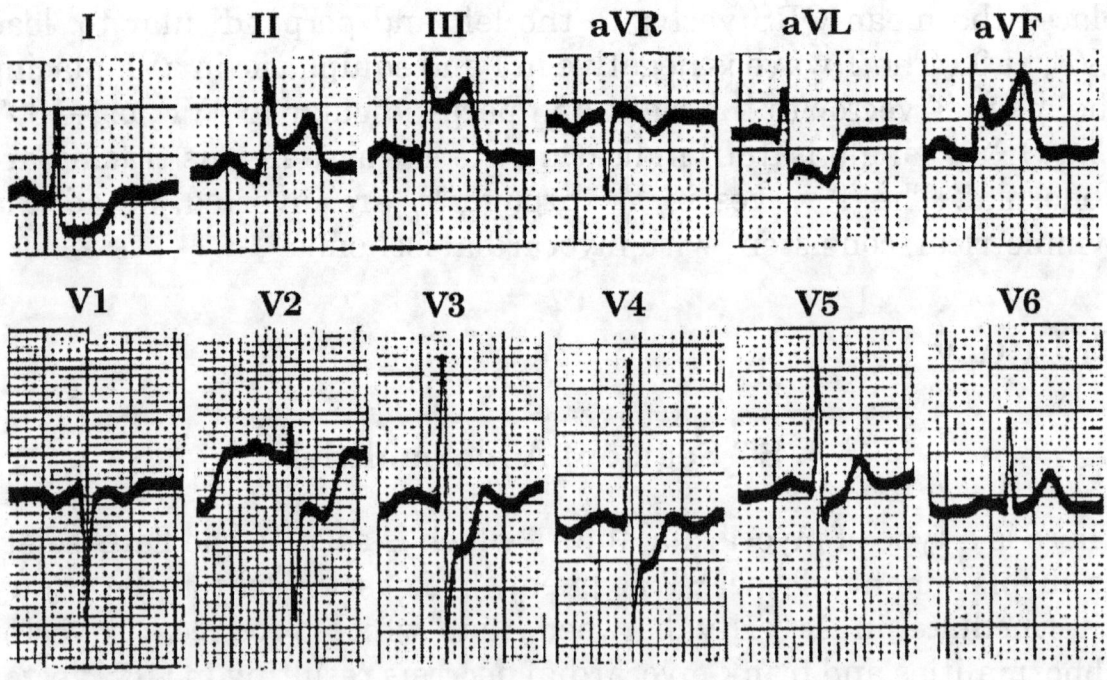

Example 4

In Example 4, the ST in lead I is depressed or negative and elevated or positive in lead aVF. The ST vector is, therefore, oriented to the right and inferiorly. The ST is isoelectric or zero in aVR indicating that the ST vector must be perpendicular to aVR in the right inferior quadrant, or +120°. The ST vector is pointing to the inferior wall of the left ventricle, the location of the epicardial injury.

D. Infarction

General Considerations

Myocardial infarction is the result of interruption of coronary artery blood flow causing ischemia, injury and tissue necrosis. Prior to the use of thrombolytic agents such as, streptokinase and tPA, myocardial infarction could not be reversed. However, if a thrombolytic agent is administered intravenously to a patient with an acute myocardial infarction within 2 to 4 hours after the onset of chest pain, coronary reperfusion may take place, minimizing or even eliminating tissue necrosis.

Myocardial infarctions (MI) are classified as to age, location, and presence of conduction defects. The age of the infarction is described as hyperacute, acute, indeterminate, or old. Locations are anterior, anteroseptal, anterolateral, apical, inferior, inferolateral, and true posterior. Recognition of the MI, however, is much more important than determining its location.

1. Zones

A typical transmural myocardial infarct consists of three zones (Figures 11, 12):

a. Area of ischemia which surrounds the injury region

b. Zone of injury surrounding the dead zone

c. Central region of necrosis, or dead zone

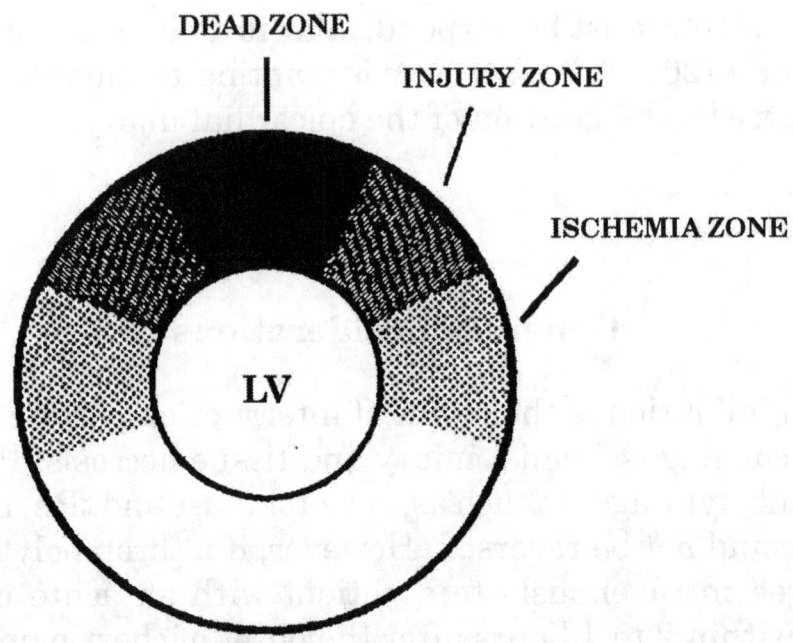

Figure 11: Cross Section Schematic

The ischemia zone accounts for T wave abnormalities. The injury zone produces ST segment abnormalities. The dead area is responsible for prolongation and redirection of initial QRS forces, primarily the initial 0.04-sec of the QRS interval.

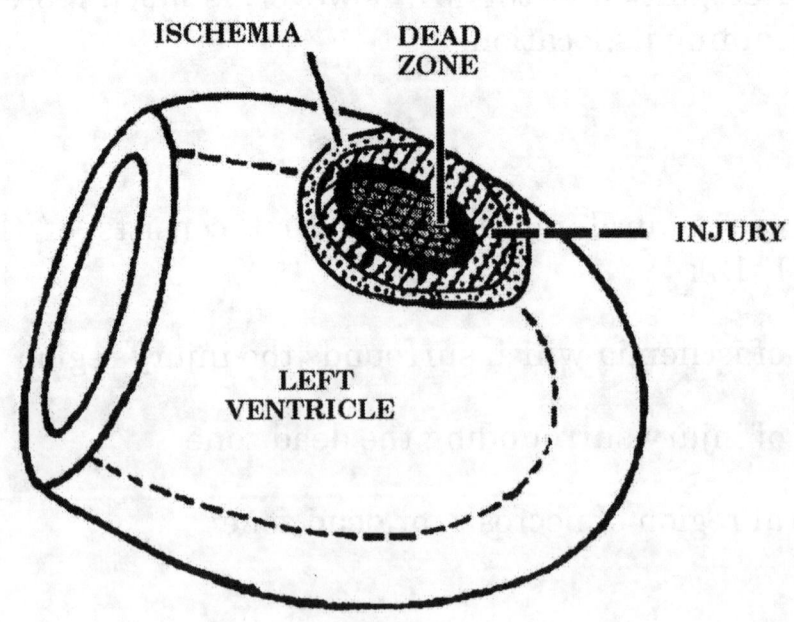

Figure 12: Surface View

Myocardial infarction may be associated with the following electrocardiographic abnormalities:

a. The direction of the initial QRS vector, up to the first 0.04-sec of the QRS complex, is changed and redirected to point away from the infarction site.

b. The mean T vector is changed in direction because of the ischemia surrounding the infarct and points away from the site of infarction. It is more or less parallel to the initial QRS vector. This means that leads with pathologic Q waves have inverted T waves, and leads with initial R waves have upright T waves.

c. The ST vector, which is due to injury, points toward the site of infarction. Since it is directed opposite to the initial QRS and T vectors, leads with elevated ST segments have Q waves and inverted T waves. Leads with depressed ST segments have initial R waves and upright T waves.

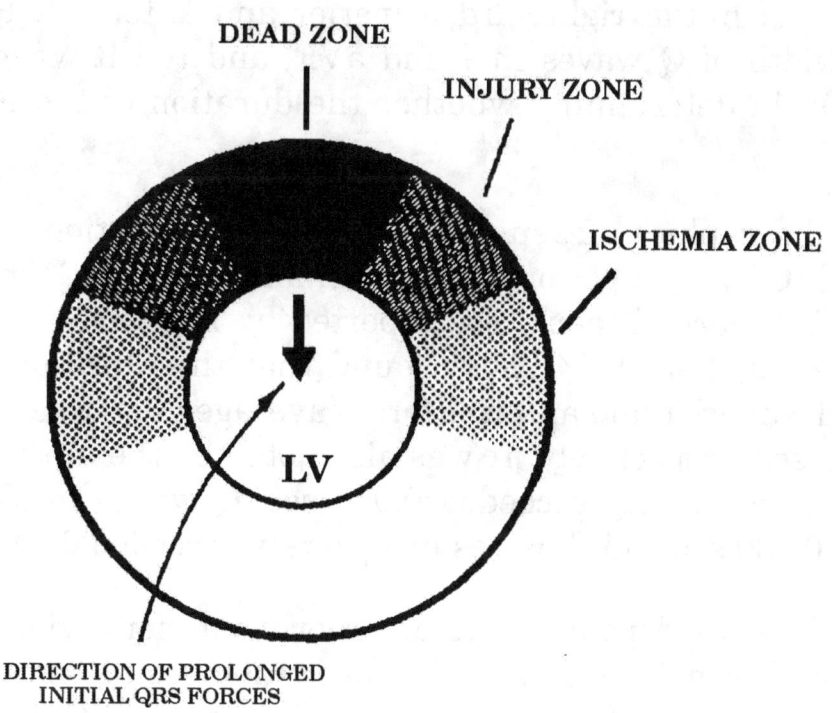

Figure 13

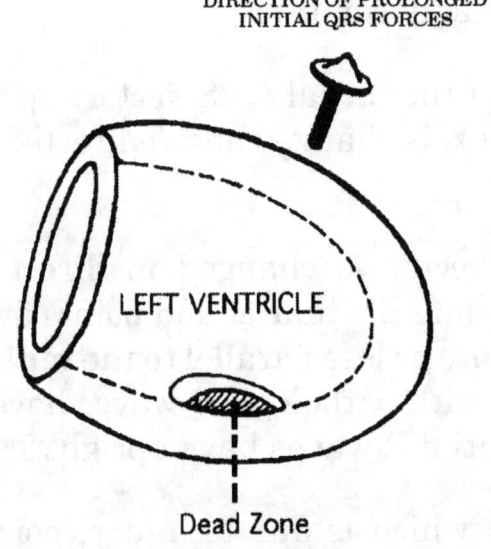

DIRECTION OF PROLONGED
INITIAL QRS FORCES

LEFT VENTRICLE

Dead Zone

Figure 14

2. Assessment of Initial QRS Forces: The 2, 3, 4 Rule

Evaluation of initial rightward, superior and anterior QRS forces, i.e., the width of Q waves in I and aVF, and the R wave in V_1 is very useful in determining whether the duration of these forces is prolonged.

Vectorcardiographic measurements of the duration of initial rightward (Q in I), superior (Q in aVF), and anterior (R in V_1) QRS forces in 510 normal men were reported by Draper and Pipberger in Circulation, Dec. 1964. They found that the duration of initial rightward, superior and anterior forces averaged 0.019 sec, 0.021 sec and 0.033 sec, respectively. It was also noted by the authors that Q waves in lead I rarely exceeded 0.020 sec, Q waves in aVF rarely exceeded 0.030 sec and R waves in V_1 rarely exceeded 0.040 sec.

The ventricles are depolarized in an orderly manner region by region (Figure 15). Normal ventricular depolarization begins on the left side of the septum moving from left to right. This causes a Q wave in lead I. Depolarization may be directed superiorly or inferiorly. When these initial forces are directed superiorly, a Q wave must be inscribed in aVF. Initial forces are normally oriented anteriorly causing an R wave to be inscribed in V_1. To summarize the above depolarization, the activation wavefront is directed to the right, superiorly and anteriorly.

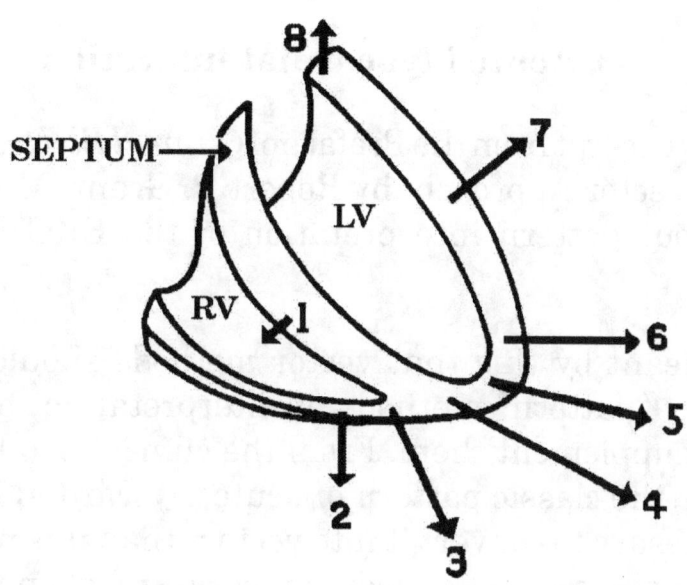

Figure 15: Sequence of Ventricular Activation

The "2, 3, 4 rule" represents one method of screening initial QRS forces. It should be made clear that this rule is a "rule of thumb" with limitations. Grant stressed the importance of the initial 0.04-sec vector as to its direction. The 2-3-4-rule considers both direction and duration of initial QRS forces as follows:

a) The *duration* of initial *rightward* QRS forces, that is, **the width of the Q in lead I**, should not exceed 0.02 sec. If it does, it could be the result of an anterior or lateral myocardial infarction.

b) The *duration* of initial *superior* QRS forces, that is, **the width of the Q in lead aVF**, should not exceed 0.03 sec. If it does, it could be the result of an inferior myocardial infarction.

c) The *duration* of initial *anterior* QRS forces, that is, **the width of the R in V_1 or V_2**, should not exceed 0.04 sec. If it does, it may be the result of a true posterior myocardial infarction or right ventricular hypertrophy.

161

Prolonged initial QRS forces are present not only in myocardial infarction but also in right ventricular hypertrophy, left ventricular hypertrophy, acute cor pulmonale (right ventricular strain), W-P-W syndrome, and left bundle branch block.

3. Acute Myocardial Infarction

The following excerpt from the Preface of *Clinical Electrocardiography*, The Spatial Vector Approach, by Robert P. Grant, M.D., conveys his thoughts about pattern interpretation of the EKG and the vector approach.

"It is not meant by this that vector methods should supplant the more familiar "pattern" methods of interpretation, but rather that they should supplement them. From the clinical point of view, when a tracing has the classic pattern of acute myocardial infarction, it is no more necessary to convert it into vectors than it is necessary to get an accurate measurement of body temperature when the patient has an obvious raging fever. However, when the tracing is perplexing or borderline, or when there is a slight difference in a follow-up tracing which is difficult to evaluate, then the vector method is the most accurate, objective, and rational method for interpretation that is so far available". This opinion of Dr. Grant is just as valid today as it was when he expressed it.

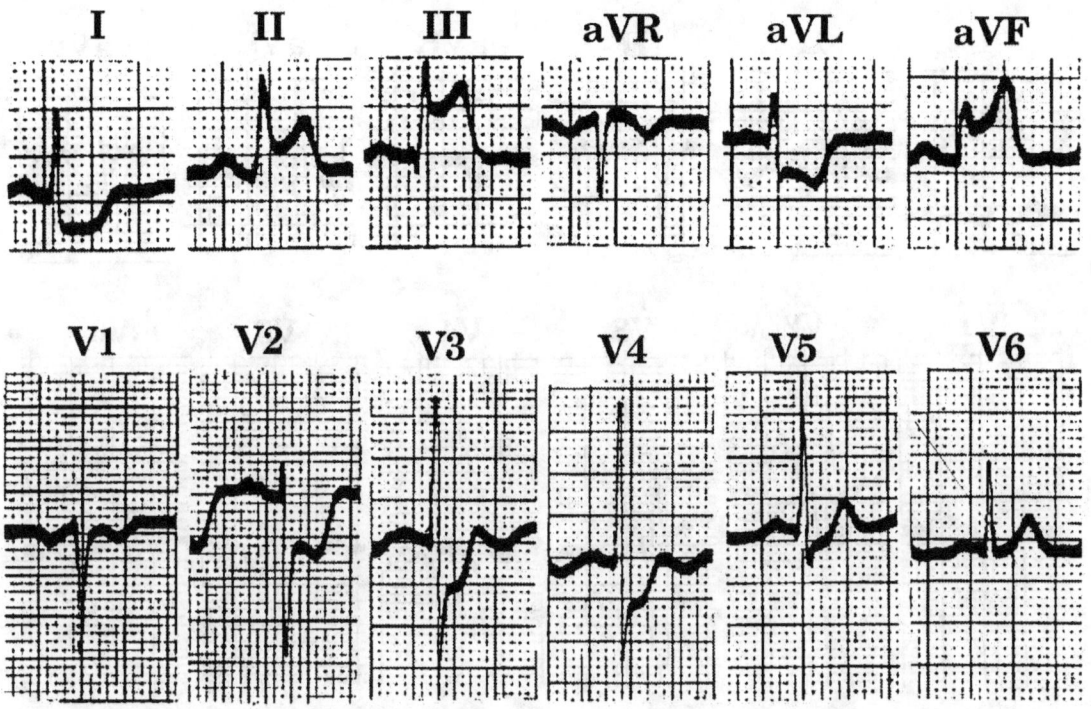

Example 5: Hyperacute Inferior MI

The EKG in Example 5 has the classic pattern of an acute inferior wall myocardial infarction. It is not necessary to determine the location of the ST vector in degrees. The ST segment displacement is obvious. There is ST elevation in leads II, III and aVF and depression in I, aVL and V_{2-5}. There are upright T waves in II, III and aVF indicating that the ST and T vectors are directed inferiorly pointing to the site of the infarction. When both ST and T vectors are oriented in the same direction, the infarction is referred to as "hyperacute".

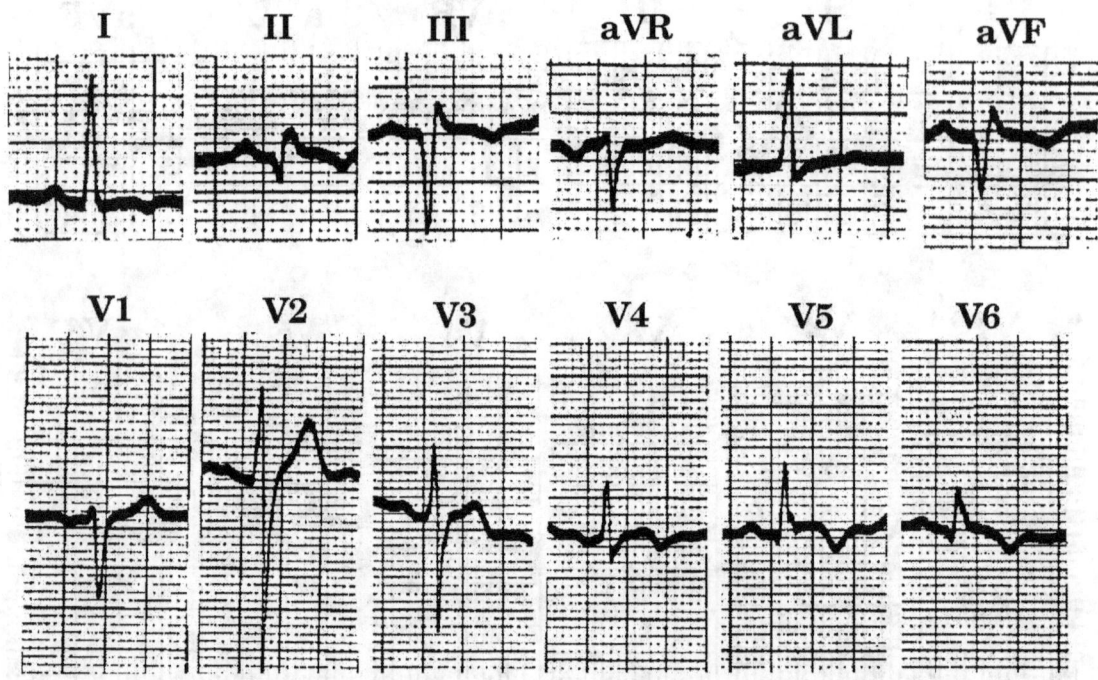

Example 6: Evolving Inferolateral MI

Example 6 was taken three days after the EKG in Example 5. Initial superior QRS forces are prolonged as indicated by the Q wave in aVF that measures more than 40 milliseconds. The "3" part of the 2-3-4 rule states that initial superior QRS forces should not exceed 30 milliseconds. The Q wave in aVF is pathologic consistent with an inferior wall myocardial infarction.

The QRS is positive in I and negative in aVF indicating that the mean QRS vector is directed to the left and superiorly between 0° and –90°. The QRS in lead II is almost equiphasic but slightly more positive than negative. This narrows the location of the QRS vector between 0° and –30°. However, the R wave in I is larger than the R wave in aVL. This places the mean QRS vector at about –5°. The T wave in Example 6 is equiphasic in I and negative in aVF placing the T vector at –90°. The QRS-T angle is 95°, which is abnormal. ST segment elevation present in V_4, V_5, and V_6 along with T wave inversion indicates extension of the myocardial infarction to the lateral wall of the left ventricle.

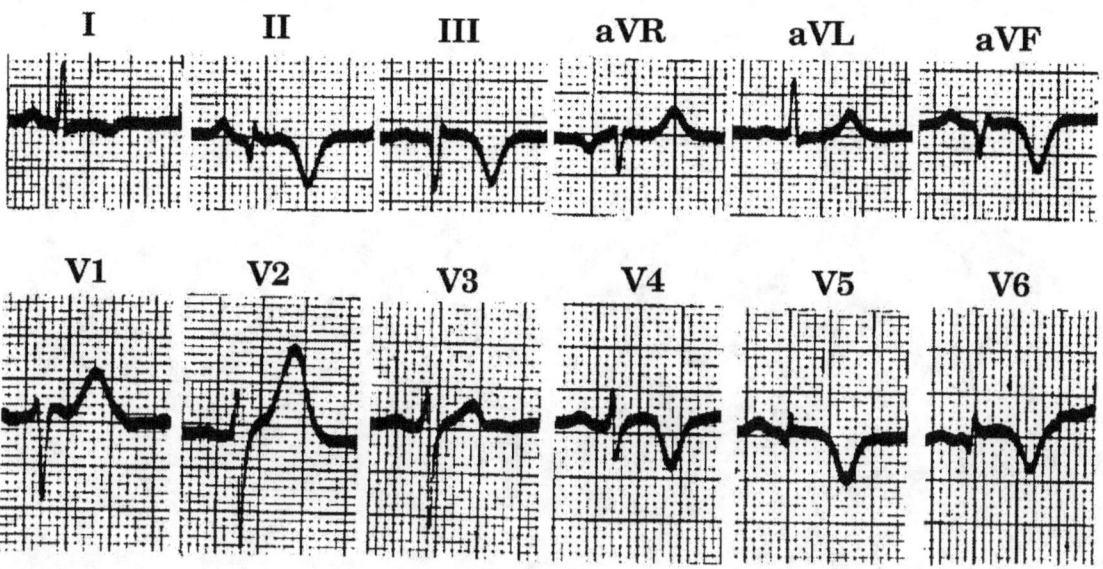

Example 7: Evolving Inferolateral MI

Example 7 was taken six days after Example 6 and ten days after Example 5: it reveals deep and symmetrically inverted T waves in II, III, aVF, and V_{4-6}. The T is equiphasic in I and negative in aVF placing the T vector at –90° This T vector has the same location as in Example 6, but its magnitude is greater.

4. Spatial Interpretation

To assure an accurate EKG interpretation, the following information must be obtained:

 a. Location of the mean QRS vector

 b. Location of the mean T wave vector

 c. QRS—T angle

 d. Location of the mean P wave vector

 e. Duration of initial QRS forces (2-3-4 Rule)

CHAPTER SEVEN:
VENTRICULAR PRE-EXCITATION

Wolff-Parkinson-White Syndrome

1. Historical Considerations

This abnormality, most likely a congenital disorder, affects people of all ages. Most individuals with the condition have no associated heart disease. However, it may occur in the presence of certain congenital heart disorders and thyrotoxicosis. It is clinically important because of a clear-cut association with paroxysmal tachycardia. It is often not apparent because routine electrocardiograms are frequently normal.

The diagnosis of ventricular pre-excitation depends on certain electrocardiographic criteria. Its presence, however, should be suspected when a patient gives a history of periodic episodes of fast irregular heartbeat.

Wilson, in 1915, reported a case of a patient who exhibited the characteristics now associated with pre-excitation syndrome, namely, intermittent anomalous atrioventricular conduction and episodes of a fast heartbeat. He thought it represented bundle branch block.

It was first recognized as a distinct clinical entity by Louis Wolff, John Parkinson, and Paul D. White who also believed it to be bundle branch block. Details on 11 cases with this anomaly were published by them in 1930 in a communication entitled, *Bundle Branch Block with Short PR Interval in Healthy Young People Prone to Paroxysmal Tachycardia* (Amer. Heart J., Vol V, No 6, August,1930). The pre-excitation disorder has since been referred to by the eponym "Wolff-Parkinson-White" (WPW) syndrome.

The classic form of ventricular pre-excitation is characterized by a short PR interval, abnormally wide QRS complex, and ST-T wave abnormalities. These findings are very suggestive of bundle branch block. It is understandable why this abnormality was mistaken for a conduction defect. It has been well established, however, that WPW syndrome is clearly *not* a bundle branch block.

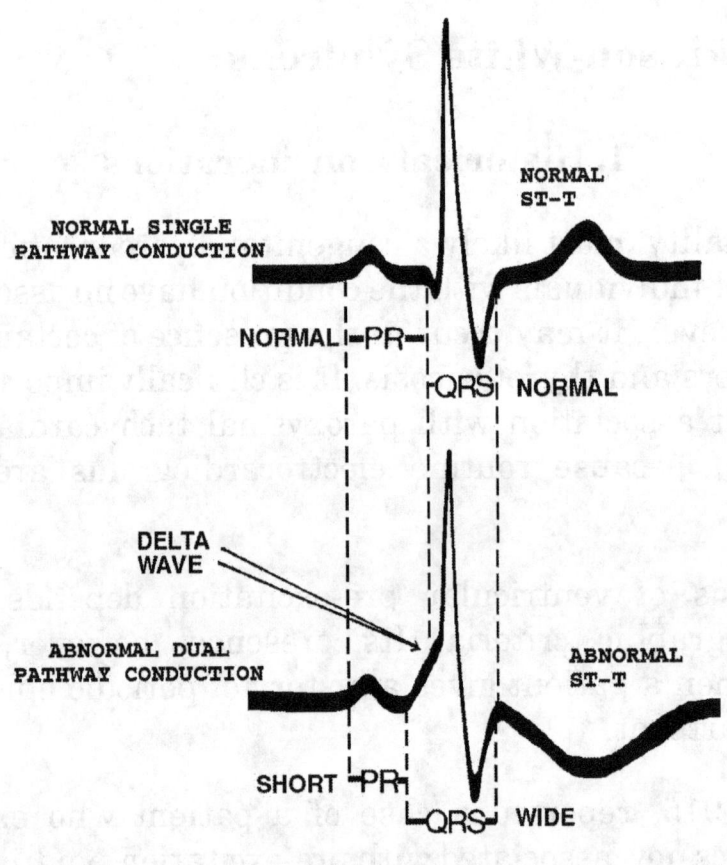

Figure 1

In Figure 1, a normal cardiac cycle is contrasted with that of ventricular pre-excitation. In the lower illustration, note the delta wave, the short PR interval and wide QRS complex. The ST-T wave abnormalities are due to abnormal depolarization of the ventricles.

2. Etiology

Several theories have been proposed to explain the mechanism of ventricular pre-excitation. Although proponents of the various theories may disagree about the mechanism, all agree that premature activation of one or other ventricle is the cause of the electrocardiographic abnormalities. It is also generally accepted that premature depolarization involves only a small portion of the ventricular myocardium, while the remaining ventricular muscle is activated in the normal way.

The concept most widely accepted is the "dual pathway" theory. It postulates that there is an accessory atrioventricular tract that permits an impulse to bypass the AV and activate a portion of ventricular myocardium. The bypass pathway impulse excites the ventricle prematurely producing a delta wave and short PR interval. The remainder of the QRS complex that is inscribed after the delta wave is produced by the normal pathway impulse. (Figure 2, A & B).

A lateral accessory bypass tract has been identified at post mortem examination in about 50% of patients with WPW syndrome. This finding supports the dual pathway theory.

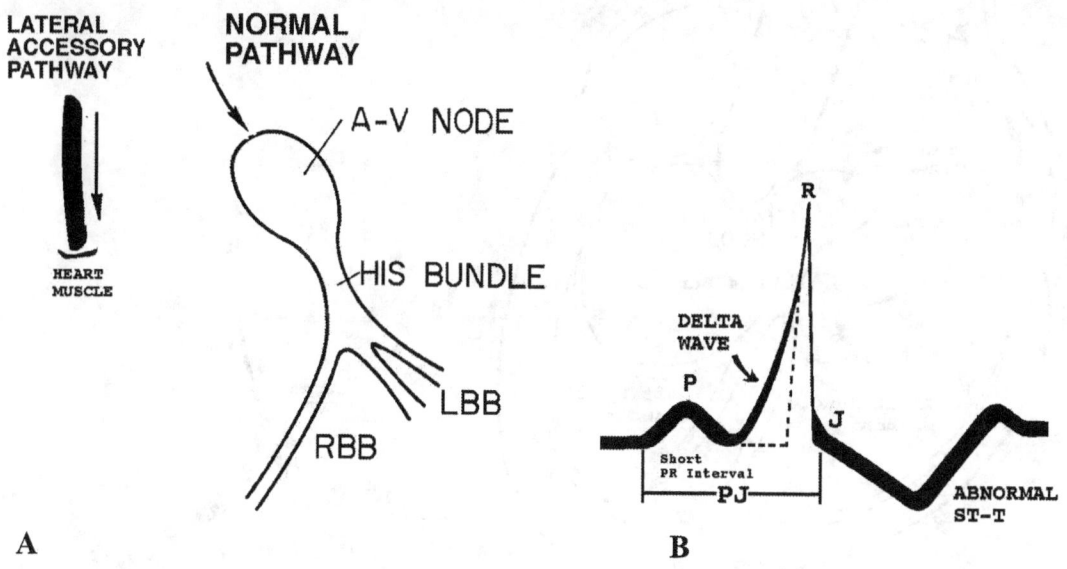

Figure 2

In Figure 2A, an accessory bundle of specialized conducting tissue is shown entering the ventricular muscle remote from the normal pathway. An initial deflection referred to as the delta wave demonstrates slow muscle conduction. A short PR interval is produced because of encroachment by the delta wave (Figure 2B).

The dotted line in Figure 2B outlines a normal PR interval and QRS complex. The duration of the QRS complex is prolonged at the expense of the PR interval, i.e., the wider the QRS, the shorter the PR interval. The letter "J" refers to the point where the QRS complex ends and the ST segment begins. The interval between the beginning of the P wave and the end of the QRS complex is called the P-J interval. Its duration remains virtually constant during normal and anomalous AV conduction. It has been stated in an earlier chapter that abnormal ventricular depolarization results in abnormal ventricular repolarization. One of the characteristics of WPW is abnormal ventricular depolarization. It follows that there will be ST-T wave abnormalities (Figure 2B).

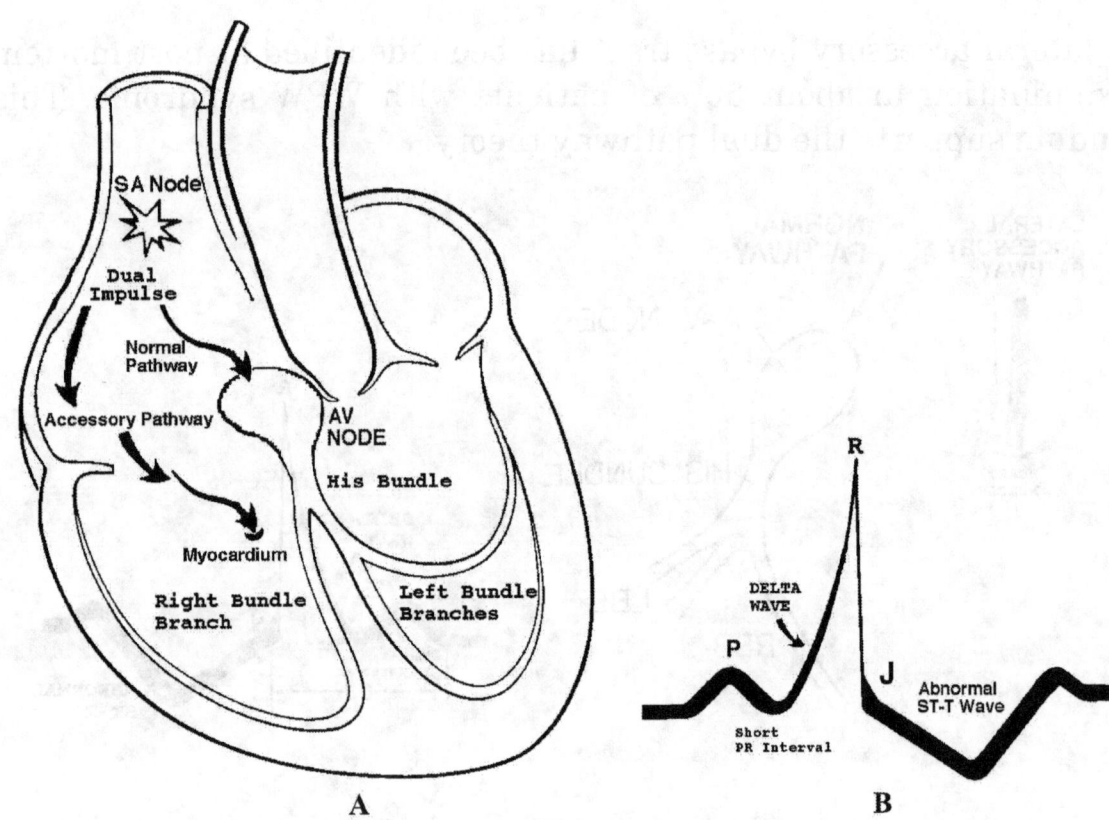

Figure 3: Dual Pathway

In Figure 3A, the impulse from the SA node is shown divided into two separate activation fronts, one moving through the normal pathway and the other along the accessory pathway. The accessory pathway impulse, moving more rapidly than the normal pathway impulse, reaches ventricular muscle sooner and produces the short PR interval. A portion of the involved ventricle is activated prematurely by the bypass pathway impulse and accounts for the delta wave. Once the bypass pathway impulse reaches the ventricle, subsequent conduction is much slower because heart muscle is a poor conductor (Figure 3B).

The normal pathway impulse proceeds at a slower rate than the bypass pathway impulse until it passes through the AV node. At this point conduction is much faster because of the specialized conduction tissue. The normal pathway impulse, eventually catching up with and overtaking the bypass pathway impulse, activates the rest of the ventricular myocardium.

The QRS complex of WPW syndrome is a type of fusion complex, the result of a single impulse. The initial portion, the delta wave, is recorded because of pre-excitation of one of the ventricles. The remainder of the QRS is inscribed in the normal way because its impulse is conducted to the myocardium through the bundle branches and Purkinje fibers. This is in contrast to the more common ventricular fusion complex that is the result of *two* impulses, one coming from the atrium and the other from the ventricle.

WPW syndrome may simulate the EKG abnormalities associated with

a. Myocardial infarction

b. Ventricular hypertrophy

c. Bundle branch block

3. Types of WPW Syndrome

a. Type A

In type A WPW syndrome, ventricular pre-excitation begins in the left ventricle and the delta wave vector is directed anteriorly (Figure 4). The delta wave in V_1 will be positive.

171

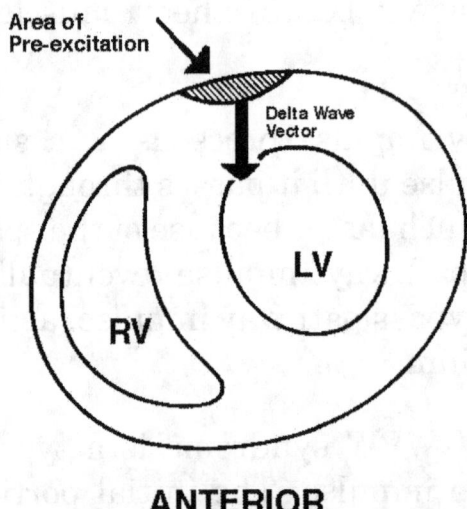

Figure 4: Type A

b. Type B

In type B WPW syndrome, ventricular pre-excitation begins in the right ventricle and the delta wave vector is directed posteriorly (Figure 5). The delta wave in V_1 will be negative.

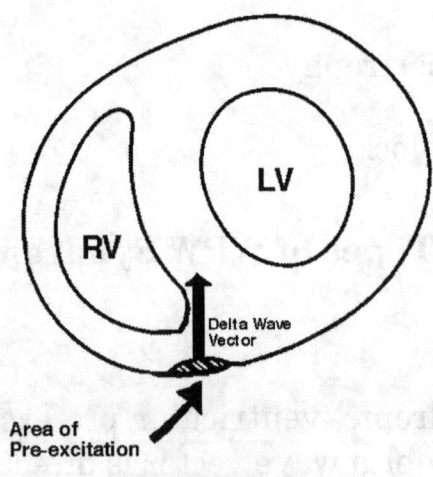

Figure 5: Type B

B. Lown-Ganong-Levine Syndrome (LGL)

LGL is characterized by a short PR interval, a normal QRS complex, normal ST segment and T wave, and frequent episodes of paroxysmal tachycardia. It occurs predominantly in middle-aged women in the absence of organic heart disease. Some authorities consider LGL syndrome a variant of WPW.

When LGL conduction occurs, there is no *direct* premature activation of the ventricle. The accesory pathway bypasses the AV node and re-enters the conduction system at a point distal to the AV junction. This produces a short PR interval but a normal QRS complex (Figure 6A).

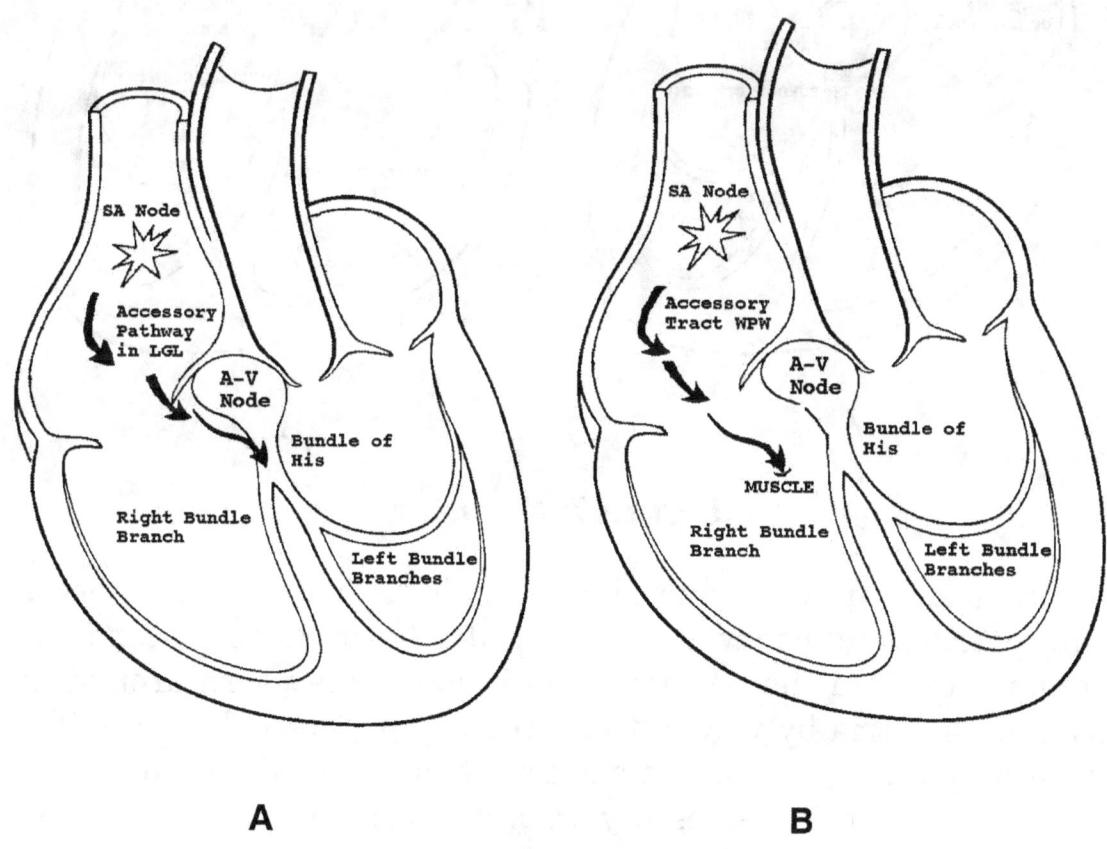

A   B

Figure 6

The above figures illustrate LGL and WPW bypass tracts. The difference between LGL syndrome and WPW syndrome is that in the former, the bypass tract impulse is conducted to the bundle of His (Figure 6A), whereas in the latter, the bypass tract impulse is conducted to ventricular muscle (Figure 6B). There is no regional ventricular pre-excitation in LGL syndrome, and therefore, no delta wave as in WPW.

C. Re-entry Mechanism

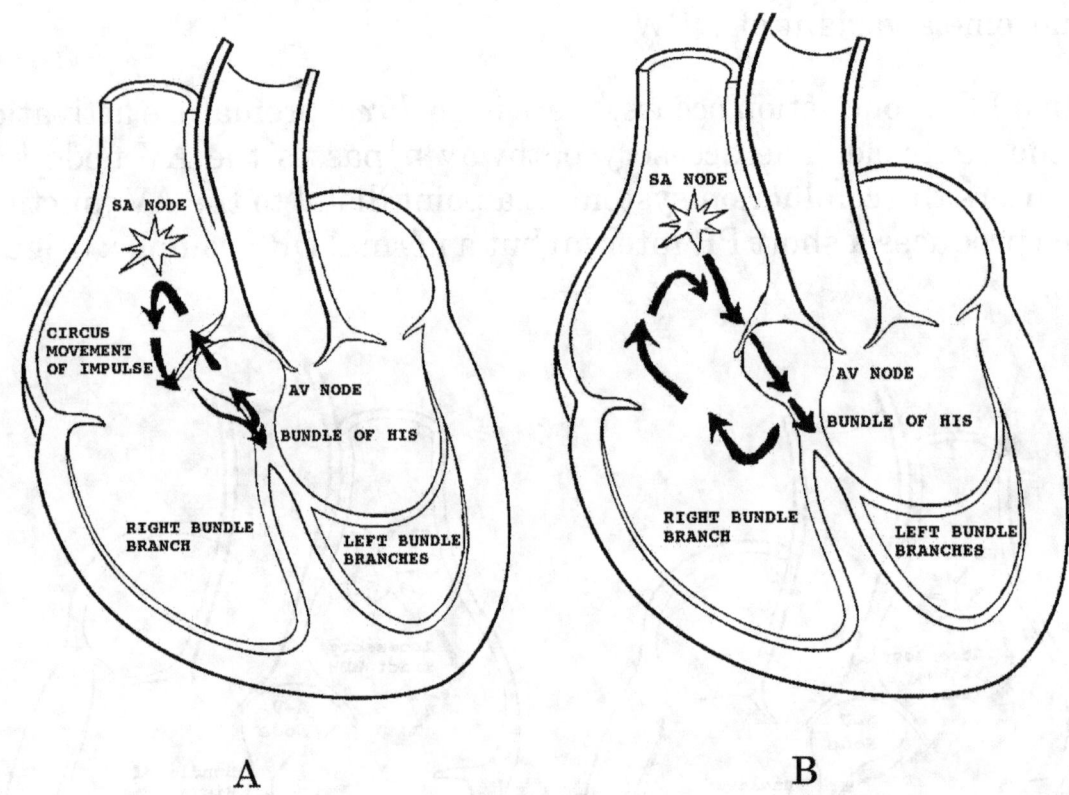

A   B

Figure 7: Re-entry

Bypass tracts may conduct impulses in two directions, forward (anterograde) and backward (retrograde) An impulse descending to the ventricles via the AV node and junction may turn around and re-enter the atria by way of the accessory pathway, but only if the accessory pathway is conducting in a retrograde fashion. An impulse could move down the accessory tract and turn around and re-enter the atria via the AV node, but only if the AV node is conducting in a retrograde fashion. In either case, re-entry and re-excitation of the atria produce a rapid reciprocating arrhythmia. What is difficult to explain is why there are times when a given tract does not conduct in one or other direction.

Figure 7A illustrates an impulse moving from the SA node to the AV node through the accessory bypass tract and re-entering the atria by way of the AV node. Figure 7B illustrates an impulse descending to the AV node through the normal pathway and re-entering the atria via the accessory bypass tract.

The diagnosis of WPW syndrome is dependent entirely on the presence of the delta wave. If a delta wave cannot be identified, it is impossible to make the diagnosis. Like any electrocardiographic deflection, a delta wave has duration, amplitude and direction. The duration of the delta wave is usually 0.04 to 0.06 sec and its amplitude rarely exceeds 5 mm. The mean frontal delta wave vector may range between −75° to +120° but most commonly is around +15°. The delta wave vector may be directed anteriorly (positive deflection in V_1) or posteriorly (negative deflection in V_1).

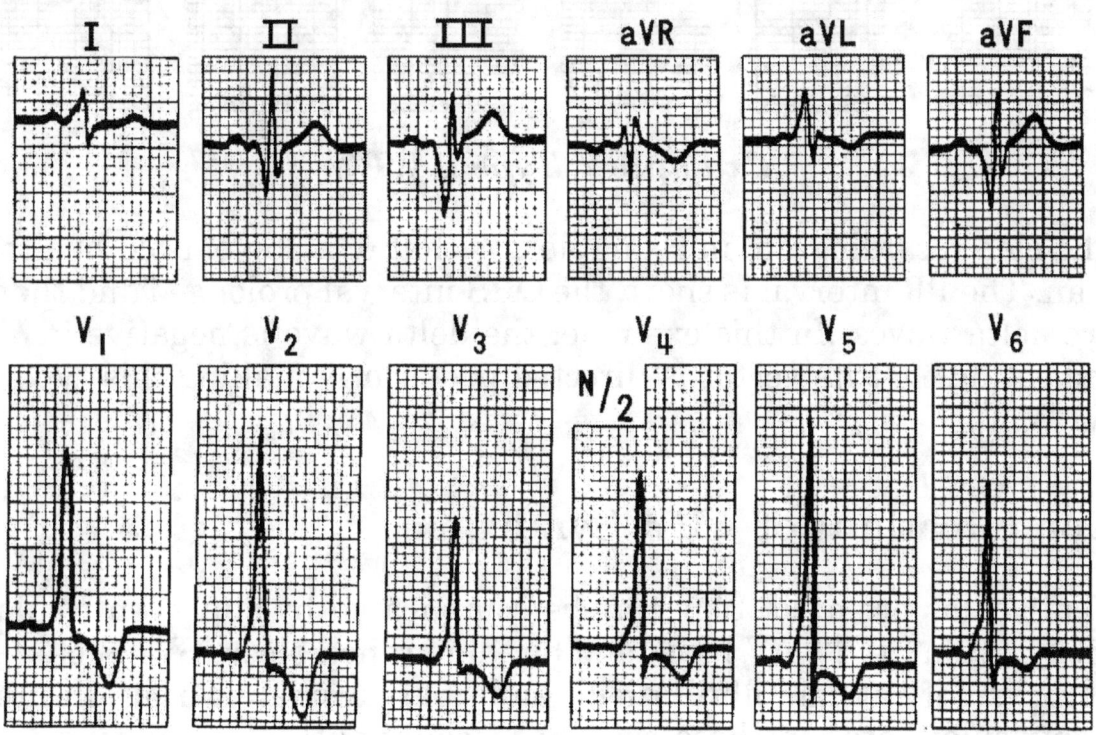

Example 1: Type A WPW

The electrocardiogram in Example 1 illustrates type A WPW syndrome. Note the short PR interval, the prolonged QRS interval, and the presence of delta waves. In this example, the delta wave is positive in V_1, indicating that its vector is directed anteriorly. Do

175

not mistake this tracing for an inferior wall MI because of the large Q waves in leads II, III, and aVF. These initial negative deflections represent the delta wave in those leads.

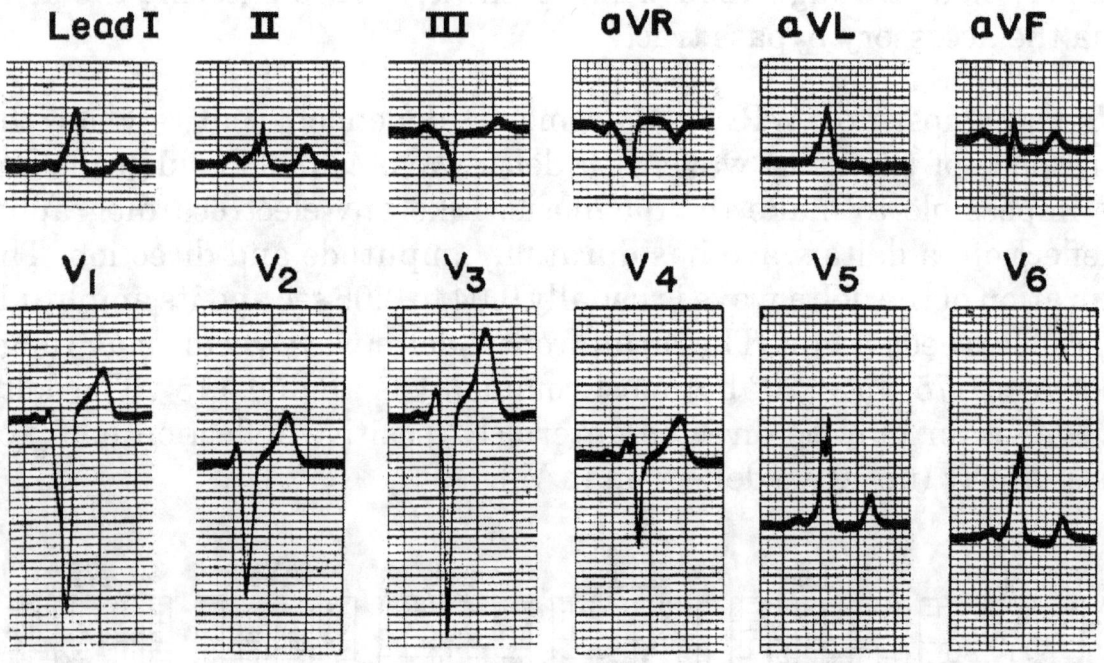

Lead I   II   III   aVR   aVL   aVF

V₁   V₂   V₃   V₄   V₅   V₆

Example 2: Type B WPW

The electrocardiogram in Example 2 is that of a 35-year-old healthy man. The PR interval is short, the QRS interval prolonged and there are delta waves. In this example, the delta wave is negative in V_1, indicating that its vector is directed posteriorly making this type B WPW.

Complications of WPW Syndrome

Ventricular pre-excitation may be complicated by supraventricular tachycardia (SVT), AV block and bundle branch block. SVT includes paroxysmal atrial fibrillation and paroxysmal atrial flutter. Ventricular rates in these cases, especially with atrial fibrillation, often exceed 200 beats per minute.

1. Atrial Fibrillation

Atrial fibrillation, the most common arrhythmia that complicates WPW syndrome, is characterized by the following:

a. Usually paroxysmal (sudden unexpected onset)

b. Associated with anomalous conduction

c. Commonly fluctuates between anomalous and normal conduction

d. Usually associated with ventricular rates exceeding 200 beats per minute

e. Very resistant to treatment with digitalis and other anti-arrhythmia medication

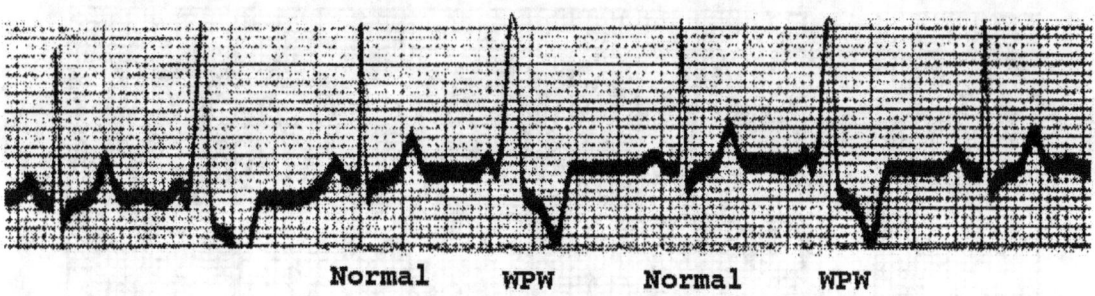

Normal   WPW   Normal   WPW

Example 3

The rhythm strip in Example 3 illustrates alternating normal and pre-excitation complexes.

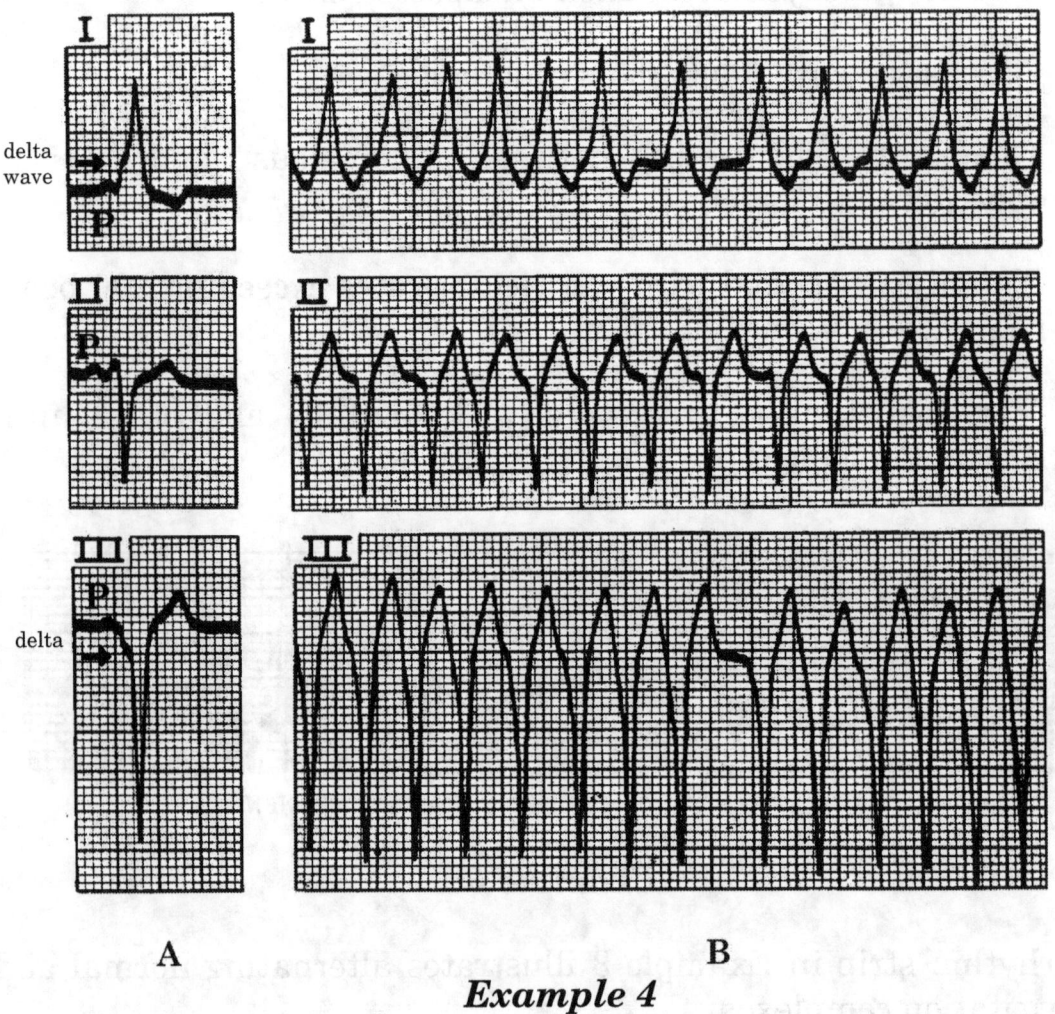

Example 4

Example 4A is from a patient diagnosed with WPW syndrome. P waves, delta waves and a short PR interval are present in all three leads. The tracing in 4B is from the same patient during an episode of tachycardia. The rhythm is atrial fibrillation with a ventricular rate of about 220. The configuration of the QRS complexes during the tachycardia is the same as present during sinus rhythm. Because abnormal QRS complexes are usually associated with atrial fibrillation of WPW syndrome, many cases have been mistakenly diagnosed as "ventricular tachycardia".

178

Treatment

Specific methods of treatment of arrhythmias associated with ventricular pre-excitation are beyond the intended scope of this manual. However, a few words about the subject are appropriate given the therapeutic advances in the treatment of this disorder.

Therapy of symptomatic supraventricular tachycardia associated with WPW syndrome includes antiarrhythmic drugs, surgery, and direct current catheter ablation of the accessory pathway(s).

The surgical approach to treatment usually requires general anesthesia and open chest visualization and transection of the bundle of His. This results in complete heart block making necessary the implantation of a permanent pacemaker. Surgical techniques have been refined and the surgeon is able to identify and transect the accessory bundle without causing heart block.

Catheter ablation using direct current (DC) energy can result in the elimination of the tract responsible for re-entry tachycardia. However, the incidence of complete heart block, the relatively high failure rate and the need for general anesthesia make this procedure less than ideal.

A newer form of catheter ablation therapy using radiofrequency (RF) energy has been shown to be a safe, effective and cost-efficient means of treating patients with reentrant tachycardia.

PATTERN-SPATIAL DICTIONARY

A. Terminology

Cardiac deflections are usually described using pattern terminology. Most instructors who teach interpretation rely more often than not on pattern descriptions of the findings present on the electrocardiogram. This manual addresses the less familiar spatial approach to interpretation of the electrocardiogram and the glossary attempts to link pattern descriptions to analogous spatial counterparts. Students of interpretation of the electrocardiogram often have difficulty converting pattern descriptions to spatial equivalents. The purpose of this dictionary is to blend pattern and spatial terminology to help the reader with the conversion.

The **"Einthoven Rule of Leads"** (term coined by author) states that a positive (upright) deflection in any lead represents forces moving *away* from the negative electrode *toward* the positive electrode of that lead. A negative (downward) deflection in any lead represents forces moving *away* from the positive electrode *toward* the negative electrode of that lead (Figure 1).

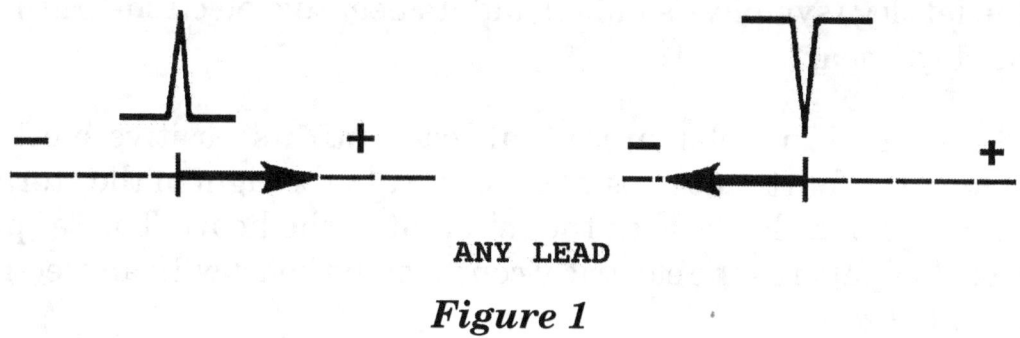

ANY LEAD

Figure 1

The three dimensions of space (3D) are described by the following lead axes.

    a. Width: left – right axis of Lead I

    b. Height: superior – inferior axis of Lead aVF

    c. Depth: anterior – posterior axis of Lead V_1

The illustration in Figure 2 is drawn to give the reader the illusion of three-dimensions.

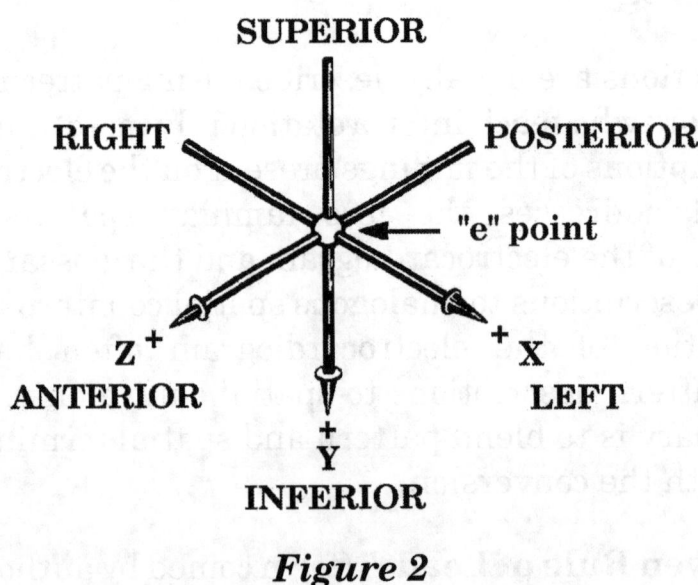

Figure 2

Lead I is a bipolar limb lead with its positive electrode located on the left arm at the shoulder and its negative electrode on the right arm at the shoulder.

Lead aVF is a unipolar limb lead with its positive electrode located on the left leg (symphysis pubis) and its negative electrode at the "e" point of the heart.

Lead $V_{1 (2)}$ is a unipolar precordial lead with its positive electrode located at the fourth intercostal space just to the right of the sternum and its negative electrode at the "e" point of the heart. The "e" point may be thought of as the electrical center of the heart with an electrical potential of zero.

A positive deflection in I represents forces moving to the left.

A positive deflection in aVF represents forces moving inferiorly.

A positive deflection in V_1 represents forces moving anteriorly.

A negative deflection in I represents forces moving to the right.

182

A negative deflection in aVF represents forces moving superiorly.

A negative deflection in V_1 represents forces moving posteriorly.

"Q waves in lead I" are equivalent to "initial rightward QRS forces". "Q waves in aVF" are the same as "initial superior QRS forces". "R waves in V_1" are equivalent to "anterior QRS forces".

Lead I: left—right axis

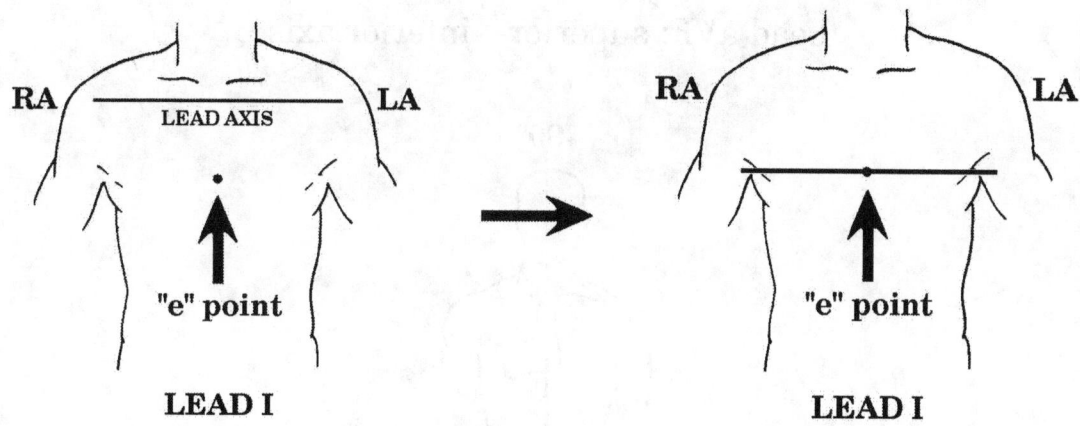

Figure 3

Q in I = *initial* rightward QRS forces

S in I = *terminal* rightward QRS forces

Lead I

R in I = leftward QRS forces

Lead aVF: superior—inferior axis

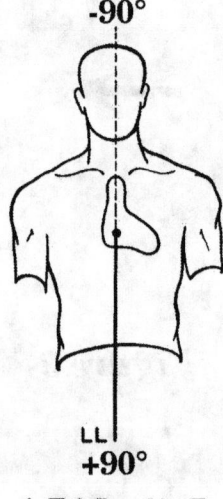

LEAD AVF

Figure 4

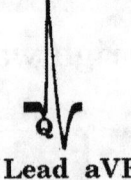

Lead aVF

Q in aVF = *initial* superior QRS forces

R in aVF = inferior QRS forces

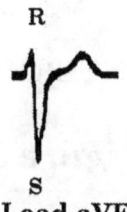

R before an S in aVF = *initial* inferior QRS forces

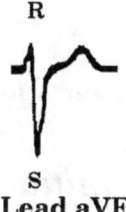

S in aVF = *terminal* superior QRS forces

Lead V_1 anterior—posterior axis

V_1, often identical in configuration to V_2, may be used as an anterior-posterior lead. Should V_1 and V_2 differ in configuration, V_1 may be more accurate as an anterior-posterior lead.

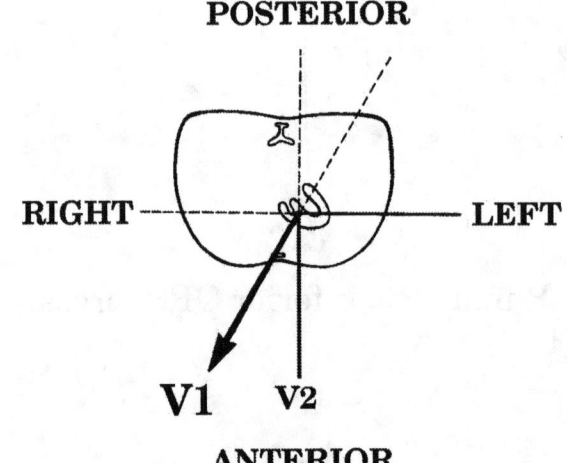

Figure 5

Lead V1

R before an S in V$_1$ = *initial anterior QRS forces*

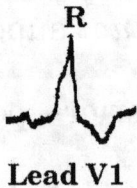

Lead V1

Wide R, no Q or S = delayed anterior QRS forces

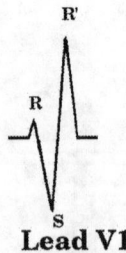

Lead V1

Wide R' of rsR' = delayed *terminal* anterior QRS forces

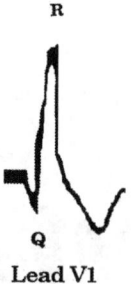

Lead V1

Wide R after a Q = delayed *terminal* anterior QRS forces

Lead V1

Wide S after an R = delayed *terminal* posterior QRS forces

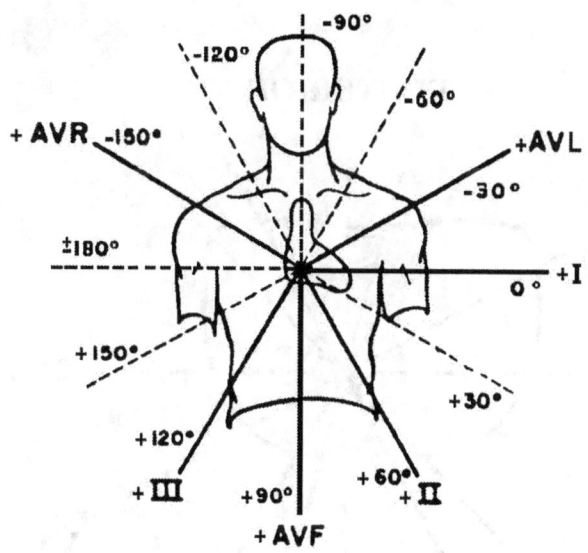

Figure 6: Hexaxial Reference Figure

The Hexaxial Reference Figure illustrated above demonstrates the bipolar and unipolar limb leads and their respective axes in degrees. When used with the Quadrant and Perpendicular Rules, it plays an important role in Spatial Analysis.

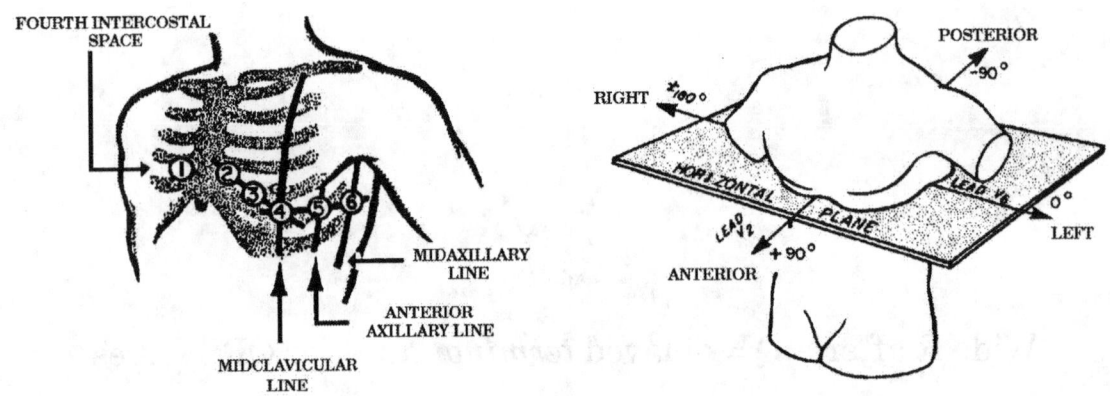

Figure 7: Chest Leads

Unipolar chest leads V_1 through V_6 are located on the anterior and lateral aspects of the chest. These locations represent the positive electrodes of the precordial leads. The negative electrode for these leads is a central terminal assumed to be located within the heart at the "e" point These electrode locations also define the horizontal plane as illustrated above (Figure 7).

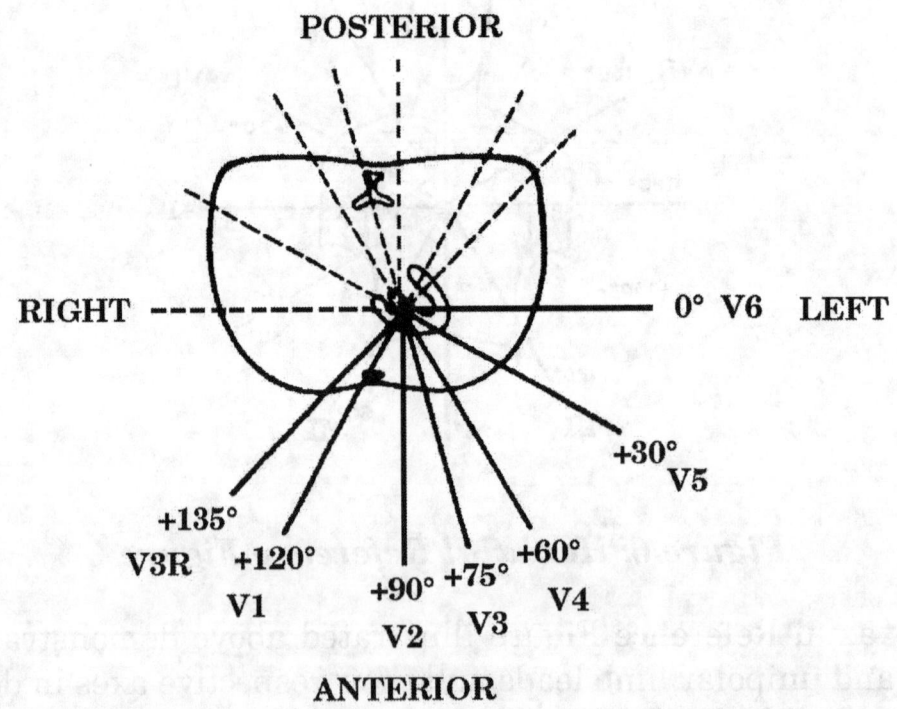

Figure 8: Horizontal (Transverse) Plane Reference Figure

188

The locations of the positive electrodes of leads V_1 through V_6 are shown in the above diagram representing a transverse section through the heart at the level of the fifth intercostal space. The axes of these leads in degrees have been added (Figure 8).

The Transverse Plane Reference Figure is useful in the determination of the anterior or posterior shift of the mean frontal QRS vector (See Chapter 3, page 84, 8. Mean Spatial QRS Vector).

B. Definitions

Anterograde conduction – Conduction in the normal forward pathway between the S-A node and ventricular myocardium.

Bifascicular Block – Block of the right bundle branch plus one of the divisions of the left bundle branch.

Block – A pathologic state in the conducting system causing the propagation of an impulse to be slowed or stopped.

Bundle Branch Block – A delay or block of conduction in the left or right branch.

Conduction – The property of impulse transmission.

Delta Wave – Initial QRS wave of slow depolarization due to a bypass tract.

Hemiblock – Block in one of the divisions of the left bundle branch.

Interval – The time between two electrocardiographic events.

Hypertrophy – Enlargement due to increase in mass.

Inverted Wave – A complex, such as P or T waves, whose major deflection is opposite in direction to what would be expected normally in a given lead.

189

Myocardial Infarction – The death of cardiac muscle due to anoxia.

P Wave – The electrocardiographic representation of atrial depolarization.

Potential – The difference in electrical charge between two points.

Pre-excitation – Early activation of a portion of the ventricles due to the presence of a bypass tract.

Purkinje Network – The terminal ramifications of the conducting system in the ventricular myocardium.

QRS Complex – The electrocardiographic representation of ventricular depolarization.

QRS-T Angle – The angle between the mean QRS vector and the mean T wave vector.

QT Interval – The duration of the time between the onset of the QRS complex (ventricular depolarization) and the end of the T wave (ventricular repolarization).

Retrograde Conduction – Conduction backward in the conducting system along a pathway from the ventricles or the A-V node to the atria.

ST Segment – The portion of the EKG between the end of the QRS complex and the beginning of the T wave.

T Wave – The electrocardiographic representation of ventricular repolarization.

Trifascicular Block – Block in the right bundle branch, one division of the left bundle branch and delayed conduction or block in the second division of the left bundle branch.

About the Author

Dr. Condo earned his Bachelor of Science degree from the University of Pittsburgh, Pittsburgh, Pennsylvania. He received his M.D. from the University of Miami School of Medicine, Miami, Florida where he completed a rotating internship. He entered general practice in Covina, California in 1962.

He did a medical residency and cardiology fellowship at Cedars-Sinai Medical Center, Los Angeles, California from 1970 to 1974 and stayed on as a Surgical Cardiologist during 1975.

After leaving Cedars Sinai, he was an invasive cardiologist at San Dimas Community Hospital, San Dimas, California and Pomona Valley Community Hospital, Pomona, California.